My Fifteen Years at IKI,
the Space Research Institute

My Fifteen Years at IKI, the Space Research Institute

Position-Sensitive Detectors
and Energetic Neutral Atoms
Behind the Iron Curtain

Mike Gruntman

Interstellar Trail Press

Library of Congress Control Number: 2022901872

Publisher: Interstellar Trail Press, Rolling Hills Estates, Calif.

1 2 3 4 5 6 7 8

ISBN: 979-8-9856687-0-4

Printed in the United States of America.

Gruntman, Mike, 1954-
  My Fifteen Years at IKI, the Space Research Institute:
  Position-Sensitive Detectors and Energetic Neutral Atoms
  Behind the Iron Curtain / Mike Gruntman.

  Includes bibliographical references and index.

Copyright © 2022 by Mike Gruntman. All rights reserved. No part of this publication may be reproduced, distributed, or transmitted, in any form or by any means, or stored in a database or retrieval system, without the prior written permission of the author.

Data and information appearing in this book are for information purposes only. Neither the author nor Interstellar Trail Press are responsible for any injury or loss or damage resulting from use or reliance, directly or indirectly, on materials in this book.

Table of Contents

Preface vii
1. From Fiztekh to IKI 1
2. Department of Space Gas Dynamics 33
3. Blazing My Own Trail 51
4. Per Aspera et cum Alcohol ad Astra 69
5. Position-Sensitive Detector: First Image and Beyond .. 85
6. Boundary Conditions 109
7. Energetic Neutral Atoms 153
8. First Steps Together 175
9. ENA Experiment That Never Flew 193
10. Neutral Atoms Reach Critical Mass 211
11. Separation of Stages in Powered Ascent 235
Appendix A. Neutral Atom Workshops in the 1980s 255
Appendix B. Brief History of ENAs 267
Appendix C. Acronyms and Abbreviations 281
Appendix D. Pronunciation Guide 285
Appendix E. Selected Bibliography 289
Index 301
Supporting materials 311

Preface

> If you ain't the lead dog, the scenery never changes.
> Lewis Grizzard
>
> If they survive, the men who go first are rarely popular with those who wait for the wind to blow.
> R. V. Jones, *Most Secret War*

Dead tired, I walked slowly through poorly lit hallways at Schiphol. The main airport in the Netherlands looked deserted on that late evening in the middle of March 1990. I felt emotionally drained. Even my fear of the immediate future had practically disappeared.

The longest day of my life was coming to an end. Several hours ago, I said goodbye to my mother and brother, who saw me off at the Moscow international airport, Sheremetyevo. I was certain that I would never see them again. Formally, this day started my four-week vacation time, which I had saved up over the previous year. Only a few people, including my immediate boss and the overseeing institute deputy director, knew that I had also submitted resignation from my position of a staff scientist in the Academy of Sciences, effective on the last day of my "vacation." My journey had begun.

Security services still controlled people's lives in the Union of Soviet Socialist Republics, USSR. The country steadily grew unstable as the Communist Party promoted a Kremlin version of a kinder and gentler "socialism with a human face." Soviet leader Mikhail S. Gorbachev relentlessly emphasized a revitalized socialist choice. He strived to save both, a disintegrating empire and his Communist Party, relying on the ossified totalitarian cadres. Gorbachev tried to reform the one and only political

Fig. P.1. Vladimir Lenin looks over a major Moscow thoroughfare today as he did in March 1990 when I left the country. Many intellectuals worldwide have been weeping for years the demise of Marxist hell in the communist Soviet Union. They still look at Lenin as an inspiration for achieving the social-justice future. Photograph (March 2018) courtesy of Mike Gruntman.

party that had ruled the country for seventy years. The Soviet people remained, well, very much Soviet, with the inertia of fear dominating the national psyche. Many intellectuals, a large part of the intelligentsia, sincerely rationalized the lack of freedom (Fig. P.1).

The terrifying moment at the Moscow airport passport control had already faded into the past. Had the officers of the border guards turned only one more page of my international passport, they would have most likely confiscated my documents, and after a brief questioning ordered me to report to a security office the next day. Then, my plan B would have been to board a train that same day and head to Warsaw in Poland with my other internal Soviet passport.[1] I counted on different bodies of the bloated repressive apparatus to communicate inefficiently and with delays in the absence of computers. A short time window should have allowed me to sneak out. The Rubicon had been crossed.

To my physics colleagues, I later compared, jokingly, the process of leaving the Soviet Union to a *tunneling transition*. In quantum mechanics, potential walls, or barriers, confine a particle such as an electron to a certain location. There is a non-zero probability,

---

1. Internal passports served as the main personal identification documents in the Soviet Union. A totalitarian state meant tight control of its subjects.

however, that the particle would manage to reach the other side of the barrier even if its energy were smaller than the barrier height. (Physicist George Gamow explained alpha decay of atomic nuclei using this counterintuitive effect in 1928.) So, one day I had been trapped in Moscow; a few days later, fate washed me ashore in California.

I am not prepared to tell the details of my complex plan for this "tunneling transition" out of the country. Suffice it to say that my colleagues and friends, unnamed here, from six countries on three continents extended their help. Nobody said no when I asked for assistance. Some contributed to my cover story and invited me to visit them as a guest in their respective countries. A few supported me en route. Others contributed to my two backup plans, B and C, which, fortunately, I did not have to try. I am truly grateful to all of them. You know who you are and a few of you, not all, are part of the story in this book. Thank you again.

Passengers spread out on an almost empty plane on the uneventful Aeroflot flight from Moscow to Schiphol. After takeoff, I saw in the window the "socialist paradise" disappearing below into the evening darkness. A melody hummed in my head. It was the unofficial student "Hymn of Aerophysicists" of my aerospace faculty:

> *Our starship ascends straight up;*
> *Small stars have lightened up in the blue sky;*
> *Wave your hand [goodbye] to the country,*
> *To a wise man in the Kremlin;*[2]
> *Because we are [proudly] rocketing on our nozzle.*

These words had taken a completely new meaning for me. "Wave your hand goodbye to a wise man in the Kremlin!" The shackles "the wise guys" put on me had been broken. I got out! Well, almost—I was still in Soviet airspace.

The bridges behind me had been burnt—I was leaving the country of my birth forever. To never turn back. I had lived in the communist Soviet Union for 35 years, the first 10 years in Central Asia,[3] including at the *Tyuratam Missile Test Range* (also known as the *Baikonur Cosmodrome*), and then 25 years in Podlipki, near Moscow (Fig. P.2). Practically all my professional life, more than 15 years, had been in the *Space Research Institute* of the USSR Academy of Sciences.

---

2. another variant of the lines: *Wave your hand [goodbye] to the Earth, To a bald man in the Kremlin*
3. Gruntman, 2004, pp. 312-315; Gruntman, 2019; Smith, 2021, p. 50

Fig. P.2. The author of this book in Podlipki near Moscow one week before his departure from the USSR in March 1990. Several close friends gathered to say goodbye and wish him luck. It was very much possible that they would never see each other again. Even these trusted friends did not know the full details of the author's "tunneling transition" plan. Photograph from collection of Mike Gruntman.

After some wandering at Schiphol, I reached the Pan American desk. It all sounds like ancient history now that Pan Am is no longer in existence. A lonely clerk searched through stacks of papers for a couple anxious (for me) minutes. He finally produced a paper ticket to Los Angeles with my name on it (see Chapter 11 for more details). Two days later I washed ashore in California—er, landed at LAX—with eighty dollars in my pocket to start a new life from scratch.

Fig. P.3. First major book about the history of IKI published in 1999.

Fast forward to 2013. My Moscow colleague Vlad Izmodenov told me about the preparation of a new book by my old Space Research Institute, widely known by its Russian abbreviation IKI,[4] to commemorate the 50th anniversary of its founding. Vlad spent one year with me at the *University of Southern California*, USC, in 1999-2000 on a postdoctoral fellowship. Today, he is a leading authority in studies of the solar system's interaction with the surrounding interstellar medium. Our science interests overlap, particularly in the area of imaging space plasmas in fluxes of *energetic neutral atoms*, or ENAs. Izmodenov suggested describing the early history of this field in the new IKI book.

Many space science missions have successfully conducted measurements of energetic neutral atoms during the last two decades. ENA instruments are operating on spacecraft today and are planned for future space missions. American, Swedish, European, and Chinese spacecraft imaged planetary magnetospheres and the heliosphere in ENA fluxes.

Very few people realize, however, that a group at IKI did some early, pioneering work on ENAs in the 1970s and 1980s. Historical publications of the institute failed to properly describe research in this field in IKI's *Department of Space Gas Dynamics*, or *Department No. 18*. Archival physics journals rarely document and publish the background of developments in science areas. (See Appendix B "Brief history of experimental study of ENAs in space.")

The first major book on the history of IKI appeared in 1999, commemorating the 35th anniversary of its founding (Fig. P.3). By that time, the head of the experimental laboratory[5] in Department No. 18, Vladas B. Leonas,

---

4. Институт космических исследований, ИКИ (Institut Kosmicheskikh Issledovanii, IKI); see *Appendix D* for a *pronunciation guide* of Russian words
5. *Laboratories* and *departments* were common administrative units in research and development organizations in the Soviet Union.

Fig. P.4. Series of books "Obratnyi Otschet..." ("Countdown...") with recollections of IKI scientists and engineers published (left to right) in 2006, 2010, 2015, and 2016.

had passed away. The publication highlighted the work of the other, theoretical laboratory of the department and said very little about energetic neutral atoms in space plasmas.[6] The book did not touch upon other notable experimental accomplishments, such as the development of the first ultrasensitive position-sensitive detectors in the country. It even sloppily listed my personal information with factual errors.

Two other books with recollections by IKI scientists (Fig. P.4) appeared in 2006 and 2010 on the occasions of the 40th and 45th anniversaries, respectively, of the institute.[7] Again, they never mentioned experimental ENA work in the Department of Space Gas Dynamics. In addition, senior IKI researchers and managers, such as heads of departments, laboratories, and groups, wrote most chapters in these publications. Nowhere could one find a perspective from the trenches, a perspective substantially detached from internal institutional politics and sniping among IKI's leading lights with often well-earned egos.

Vlad Izmodenov's suggestion to write a chapter appealed to me, as I could certainly offer such a complementary view. Besides, I had recently[8] given an hour-long lecture on the early history of energetic neutral atoms, including work at IKI, at a festive science team meeting of NASA's *Interstellar Boundary Explorer* space mission, or IBEX, in Chicago. By that time, IBEX had completed the first year of observations and produced the very first maps of the interstellar boundary of the solar system in energetic neutral atom fluxes. *Science* magazine highlighted the achievement on its

---

6. Galeev and Tamkovich, 1999, pp. 68–72
7. Obratnyi Otschet Vremeni, 2006; Obratnyi Otschet... 2, 2010
8. November 4, 2009

cover. It was the culmination of a 30-year journey, started at IKI in the 1970s, to detect these heliospheric ENAs.

IKI director (from 2001-2016) Lev M. Zelenyi and I had known each other for 40 years. Six years older than I, Lev had already been an "old-timer" working on his Ph.D. at IKI when I came in as an undergraduate third-year student in 1973. Zelenyi responded affirmatively to my inquiry about contributing to the new book.

In early 2015 I sent the manuscript in. IKI promptly requested to delete the part describing the role of alcohol in operations and the everyday life of the institute. Otherwise, they would not publish it.

Left without choice, I had to agree to this institutional censorship. In my view, the editors made a mistake insisting on eliminating stories about alcohol. The accomplishments of IKI are even more impressive when one realizes the bizarre conditions under which scientists and engineers worked in the old Soviet days. No alcohol censorship in this publication.

IKI published the book of recollections "Obratnyi Otschet ... 3" ("Countdown ... 3") with my chapter[9] in the fall of 2015. Another institute's historical book[10] followed in 2016 (Fig. P.4).

This publication significantly expands the scope of my chapter in the IKI book and includes numerous new details, stories, events, and observations. Its title "My Fifteen Year" emphasizes the word "My." This is a personal story of new, emerging fields and about the events that affected me, what I did, how and why I acted the way I did, the people with whom I interacted, and how I saw the developments. Therefore, the pronoun "I" is prominent. The book also describes consequential constraints: the "boundary conditions" on professional life imposed by the political system.

My story took place during the Cold War in the 1970s and 1980s, with the "Iron Curtain" in the subtitle. It was a time of epic struggle between the dark forces of Communism, supported by the fifth column of the international left and other collaborators in the West, and the free world led by the United States. Then, America served as inspiration for a better future and freedom to the oppressed—"a shining city on a hill." And the free world ultimately prevailed in the twentieth century.

The recent rise of the totalitarian neo-Marxist far left in the United States supports a dangerous slide into socialist hell, away from common sense. Perhaps the most destructive and dehumanizing ideology in history, socialism understands "equality" as being "identical" ("identity" in the mathematical sense), views people as members of groups rather than indi-

---

9. Obratnyi Otschet ... 3, 2015, pp. 236-261
10. Obratnyi Otschet ... 4, 2016

viduals, and relies on the coercive power of societal structures, particularly of a state, to advance its cause.

This sinister turn, if not reversed, will lead to an inevitable economic, cultural, and scientific decline and disintegration of the most successful political system founded on the unalienable rights of life, liberty, and the pursuit of happiness. How this calamity would end is anyone's guess.

The reader will find in this book glimpses of the realities of everyday life in a mature socialist society of the Soviet past. They provide not only the background to the described events but also a warning to today's generations about where intolerant neo-Marxists are dragging them into.

The book also includes explanations of life in the Soviet Union that are necessary for understanding the story. The structure and organization of the Soviet state, economy, and science did not resemble those in the market economies. The functions of the USSR Academy of Sciences also differed from those of the national academies in the United States. In the Soviet Union, the Academy of Sciences operated a vast network of well-funded research institutes in natural sciences. One could view them as analogs of the U.S. national laboratories and federally funded research and development centers or research institutes of the *Max Planck Society* in Germany.

Even the best Soviet universities were on average second-rate in science compared to the USSR Academy of Sciences. The institutes of the Academy and other non-university research establishments in nuclear, aerospace, radar, and other defense-related areas dominated fundamental and applied science and engineering. These organizations also had their internal programs for awarding Ph.D. degrees, with the quality often superior to those earned in universities and other institutions of higher learning.

Satellite photographs, government documents, personal names and organizations, foreign language features, appendices

Reconnaissance satellite photographs

Several photograph captions identify optical systems of U.S. reconnaissance satellites. The Keyhole (KH) space optical intelligence systems operated the KH-4 (Corona) cameras from 1963-1972. The KH-9 (Hexagon) cameras provided images from 1971-1984.[11] These optical reconnaissance systems relied on film returned to Earth for processing.

The satellites took photographs of the areas on the ground at various viewing angles. I did not attempt to correct distortions in oblique images

---

11. Ruffner, 1995; Gruntman, 2004, Chapter 16; Berkowitz, 2011; Pressel, 2013

that could be significant.[12] Consequently, the shown scales and compass directions are approximate.

Geographic latitudes and longitudes of particular sites in satellite photographs are given with an accuracy of one-tenth of one minute (six seconds), $0.1' = 6''$. On the ground, it corresponds to a spatial uncertainty of 185 m in the meridional north-south direction and 100-135 m, depending on latitude, in the azimuthal east-west direction. Listing more precise geographic coordinates would have required identification of the exact part of a particular site, for example, the wing of a building. Such precision is excessive and distracting.

Government materials

All used, referred to, and quoted U.S. government documents and materials are declassified and in the public domain.

Transliteration (Romanization)

Several widely accepted systems transliterate Russian words, written using Cyrillic, into the English language.[13] In this book, I use transliteration (Romanization) that largely follows traditional approaches.

Personal names

According to the Russian tradition, the name of a person consists of the first (given) name, the patronymic name, and the last name or surname (family name). The patronymic is a modified form of a genitive case of the father's name of the person. So, people usually address somebody or describe him or her using the first name followed by the patronymic. Among peers, close colleagues, and friends, many often use first names only. In addition, diminutive forms of the first names, for example, Misha, Sasha, and Volodya for Mikhail (Michael), Aleksandr (Alexander), and Vladimir, respectively, are common. In the book, I use only first and last names in the Westernized fashion.

Names of organizations

I capitalize all words in the names of organizations in the Soviet Union. In the Russian language, one capitalizes only the first word.

Umlauts

The German words with umlauts are rendered by adding the extra *e*.

Diacritics

Polish names and words show letters with diacritics when introduced. Then, these marks are dropped.

12. Figure 5.13 shows an example of such distortions.
13. e.g., Timberlake, 2004, pp. 24-27

Ukrainian cities

I render the names of Ukrainian cities from the Russian language as in the old Soviet days. When used the first time, transliteration from Ukrainian follows in parenthesis, for example, Kiev (Kyiv).

Appendices

Appendix A lists participants and titles of presentations at the first four annual international workshops on neutral atoms in space from 1984-1987. Appendix B is the section "Brief history of experimental study of ENAs in space" from a review article[14] published in a physics journal in 1997. (The American Institute of Physics Publishing authorized its inclusion.) Appendix C summarizes acronyms and abbreviations in the book. Appendix D provides a pronunciation guide for some Russian words and names.

Appendix E contains a selected bibliography with more than 180 titles. Many references to my publications could have been omitted. They are included for the sole purpose of validating the factual accuracy and timing of the events. The index, with more than 600 entries (including more than 160 individuals), follows the appendices.

The opinions expressed in the book are solely mine and not necessarily shared by my former or current colleagues and friends. Needless to say, I take the responsibility for all errors.

Mike Gruntman
California
January 2022

---

14. Gruntman, 1997a

1. From Fiztekh to IKI

Fateful telephone call

In early June 1977, my science advisor Vladas B. Leonas called me home. A few weeks earlier, I'd graduated from the *Faculty of Aerophysics and Space Research* of the *Moscow Physical-Technical Institute* with a degree of engineer-physicist after six years of studies. I was 22 years old and had one publication as the lead author (with one coauthor) in a major peer-reviewed physics instrumentation journal. My degree roughly equaled that of a master of science in the United States.

The Russian abbreviations FAKI (ФАКИ) and MFTI (МФТИ) stood for the names of my faculty and institute, respectively. People commonly called the latter *Fiztekh*. (Today the institute's name is often rendered in English as the *Moscow Institute of Physics and Technology* or MIPT.) Arguably the most elite school in physics and applied physics in the Soviet Union, Fiztekh boasted exceptionally difficult and competitive entrance examinations.

The institute graduated half a thousand physicists each year who went to work in leading research and development organizations of the Academy of Sciences and industry. Many specialized in areas of importance for defense programs. World-class Soviet physicists taught and mentored MFTI students.

Studying in Fiztekh was particularly challenging. In their secondary school years, many incoming freshmen won national, regional, and local competitions, known as "olympiads," in science and mathematics. (Secondary schools usually combined the primary, middle, and high school levels common in the United States.) Some studied in magnet-type schools with

1. From Fiztekh to IKI — Mike Gruntman

Fig. 1.1. The author of this book in a secondary school in the late 1960s. From collection of Mike Gruntman.

a focus on natural sciences. Quite a few students had spent one or two years in science boarding schools, one in Moscow and the other in Akademgorodok near Novosibirsk, established for specially gifted boys and girls.

There were no magnet-type schools in my town. So, I went to a regular secondary school where science and mathematics classes were not particularly challenging (Fig. 1.1). Fortunately, my wise teachers did not bother me too much and let me study on my own beyond the standard program.

Children of top Communist Party and government officials, the nomenklatura,[1] could get admitted into most universities and institutions of higher learning through their connections. Because of difficult coursework they rarely enrolled in Fiztekh. The privileged rather sought and entered fields of studies with a promise of coveted travel and work abroad. Some favored relatively easy educational programs followed by careers in officialdom, paved by their parents.

After graduation from Fiztekh in June 1977, I looked forward to summer vacations (Fig. 1.2) and then studies in a Ph.D. program of the Space Research Institute of the USSR Academy of Sciences. Similar to universities, such programs led to Ph.D. degrees. The fateful telephone call from my advisor altered the plans.

"Misha," began Leonas, "there is a possibility to change your situation and hire you directly to IKI as you wished it." It was true. I wanted to get a job as a staff researcher in the institute after graduation. One could write and defend a doctoral thesis and earn an indispensable Ph.D. degree

---

1. The *nomenklatura* comprised senior officials who required approval of their appointments by central national authorities of the Communist Party. The nomenklatura enjoyed numerous everyday life privileges.

while working in an academic institution. The state policies closed this avenue for me at that time as explained below. I could, however, apply to a competitive graduate study program, *aspirantura*, of the Academy of Sciences, which had a few open slots at IKI. So, I pursued that aspirantura path without much enthusiasm and gained admission.

"IKI director Roald Sagdeev just talked to me," said Leonas. "He wants to bring a young graduate from a provincial town to work with him. The guy does not have a Moscow residence permit and he also limps on both legs on the item number five. Therefore, Sagdeev cannot hire him as a staff member. The director thinks, however, that he could succeed in enrolling him into the aspirantura in your spot."

The item number five meant the fifth line in the Soviet internal passport. After the entries of a person's given name, family name, and place and date of birth, the fifth line in the passport identified a distinct "nationality" of the individual. The word nationality here meant ethnicity, such as, for example, Russian, Uzbek, Jew, Armenian, or Korean.

Fig. 1.2. A freshly minted physicist, the author of this book, on a visit to a "city of science," Akademgorodok near Novosibirsk, in July 1977 after graduation from Fiztekh and before starting his job as a staff member at IKI. From collection of Mike Gruntman.

Socialism essentially defines and treats all people as members of groups rather than individuals. This dehumanizing feature unfairly opens or cruelly closes many doors in the lives of people based on their immutable characteristics rather than on merit and "the content of their character." The Soviet government widely used nationality for advancing or discriminating various categories of people in a feat of massive inhumane social engineering. Modern neo-Marxist-Leninist proponents of demeaning affirmative action and destructive identity politics and diversity by exclusion in the American political discourse and practice, especially on university campuses, follow

1. From Fiztekh to IKI   Mike Gruntman

the steps of their inspirational Soviet counterparts. They often look, however, as pitiful imitators.

My advisor Leonas also used an idiomatic expression invoking bodily injury. In this context, limping on both legs on the item number five meant that both parents of that graduate from a provincial town were Jews. Even powerful IKI director Sagdeev had to choose his battles and could not hire such a person to work for him as a junior staff member at that particular time. It also so occurred that state antisemitism somewhat eased up the nationality "filtering" of admissions to aspirantura that summer.

A residence permit sometimes presented an insurmountable problem as well. Internal passports displayed a specific location where individuals lived. One could not change one's residence town or village at will. A totalitarian society means total tight control of the subjects. Obtaining a residence permit to live and work in Moscow would have required a major and lengthy administrative effort by IKI and would not necessarily have been successful. A permit for a graduate student for studies in aspirantura, though temporary in nature, was much easier to secure.

Leonas continued: "In contrast to the other guy, Misha, you limp only on one leg, and you live and have a residence permit in a Moscow suburb." Since the mid-1960s, I had lived near Moscow in Podlipki, also known as Kaliningrad and later Korolev (Fig. 1.3). The town was home to a major cluster of space and ballistic missile research and development organizations.[2] Leonas then concluded, "So, if you agree to vacate your aspirantura spot, Director Sagdeev thinks that he will be able to force through your hiring, per your desire, as a staff engineer to IKI against expected resistance. In one year, I will also try to change your position to a junior research fellow."

My stars had aligned! The proposition truly delighted me, and I immediately agreed.

A somber-looking unhappy head of the personnel (human resources) department of IKI tormented me for a few weeks with demands for all kinds of bureaucratic papers from various administrative bodies, more than usually required. My senior colleagues later explained his attitude: "He is in his position particularly to make sure that there are fewer people like you in the institute." As I recall, I had become a staff engineer in the laboratory headed by Vladas Leonas in Department No. 18 of IKI by early August 1977.

I provide here the dates and quote people from my memory. When I left the USSR in March 1990 (the details of that story are not for this publication) I could not take with me any notes, documents, or papers. Then I destroyed

---

2. Gruntman, 2004, pp. 279-283

practically all my personal archives since I believed that I would never be able to visit Moscow again.

Top military and civilian communist hardliners attempted to restore the totalitarian state by force in a putsch in August 1991. This coup d'état failed and the Soviet Union ultimately disintegrated. The events confirmed, however, that my anticipation of a possible restoration of the harsh regime with or without Gorbachev at the helm and resurrection of Marxism-Leninism were well-founded.

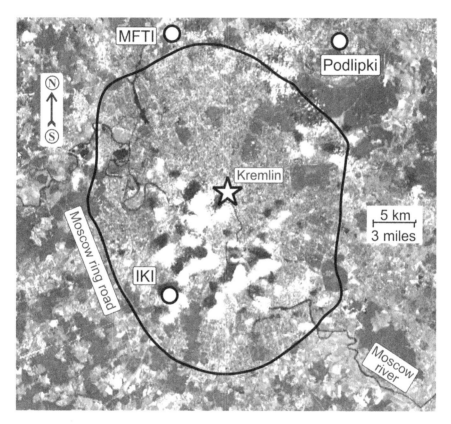

Fig. 1.3. Satellite photograph of Moscow and some locations mentioned in the book. For many years a ring road (black line), approximately 21 miles or 34 km in diameter, served as the administrative boundary of the Soviet capital. The white circles show Fiztekh (MFTI) 2.5 km north from the city, Podlipki (55°55.5' N, 37°49.0' E) 6 km to the northeast, and the Space Research Institute (IKI) in the city's southwest. Original composite satellite photograph (Landsat, mid-1990s to early 2000s) courtesy of NASA WorldWind; image processing, callouts, ring road line, and markings by Mike Gruntman.

1. From Fiztekh to IKI — Mike Gruntman

Fiztekh and Fiztekh system

My association with IKI had begun significantly earlier than joining the institute as a staff member in August 1977. It lasted fifteen years from 1973 until the late 1980s.

Bachelor of science degrees did not exist in the Soviet Union. After five or five and a half years of studies in universities or institutes, students received degrees equivalent to those of a master of science. Higher education in technical fields concentrated in specialized institutes rather than in universities that typically had programs in natural sciences but no engineering schools.

The Moscow Physical-Technical Institute stood out from other institutions of higher learning by its unique system of education and exceptional quality. In the late 1930s, several leading Soviet scientists published a letter[3] in *Pravda* (Fig. 1.4). The daily newspaper *Pravda*, the main organ of the Central Committee of the *Communist Party of the Soviet Union*, CPSU, was the most authoritative official publication in the country.

In the letter, the scientists advocated the creation of a new elite engineering school in Moscow to address a problem that no institutions in the country prepared "engineer scientists, engineer researchers, who combined the perfect knowledge in certain areas of technology with broad general education in physics and mathematics." The history of the French *École Polytechnique* that had been "established during very difficult times" of the French Revolution in 1794 to provide "real military technology and military art" inspired, in part, the proposal.[4]

After the end of World War II, the importance of a new school specifically focused on the education of physicists as an alternative to existing universities became appealing to many top officials.[5] The country poured enormous resources into the development of new weapons in the late 1940s.[6] The recent war experience and "the obvious role of science and technology in the post-war world in assuring the security of the country forced to return to the question of establishing the corresponding institution of higher learning."[7]

A former associate of Ernest Rutherford, physicist Petr (Pyotr) L. Kapitsa, and a specialist in mechanics and aerodynamics, Sergei A. Khristianovich, wrote the founding documents of the new institute. The

---

3. Muskhelishvili et al., 1938
4. Khristianovich, 1996
5. e.g., Karlov et al., 1996; Shchuka, 2012
6. e.g., Smith, 1950, p. 320; Gruntman, 2015a, p. 13
7. Karlov and Kudryavtsev, 2000

Fig. 1.4. The most important daily newspaper in the country, the official organ of the Central Committee of the Communist Party, *Pravda*, on December 4, 1938. It published a letter (Muskhelishvili et al., 1938) of nine influential scientists and engineers advocating the creation of a new elite institution of higher learning for educating research-oriented specialists in technical fields. This proposal would eventually lead to the establishment of MFTI.

The letter specifically stated that the students should be selected on a rigorous competitive basis from "the most outstanding by their abilities" graduates of secondary schools. The applicants would have to go through unique additional testing in "mathematics and physics." The letter called for the faculty and instructors of the school to consist only of the "prominent scientists who are intensely engaged in creative research work."

demands of the development of nuclear weapons, jet aviation, rocketry, missiles, radar, and electronics made the upper levels of government receptive to strengthening physics. Soviet dictator Joseph (Iosif) Stalin endorsed the effort to establish the new elite school by signing the corresponding decree on March 10, 1946.

MFTI opened its doors in the late 1940s in a Moscow suburb, Dolgoprudny (Figs. 1.3, 1.5, 1.6). Khristianovich recalled:

> It was considered first to locate [the institute] in Moscow. Here, Petr Leonidovich [Kapitsa] had played a major role [in changing the plan]. He remembered Cambridge [where he worked for over ten years with Ernest Rutherford in the Cavendish Laboratory at the University of Cambridge], a town-university, a town of colleges, where Isaac Newton walked on trails in a park on the banks of a river. A difficult program of studies was planned for students. It was necessary to have a quiet and spacious environment.[8]

Dolgoprudny was obviously no Cambridge. Accordingly, students quickly shortened the town name to Dolgopa, which also unflatteringly rhymed with a Russian word for *ass* (in the meaning of a body part, not an animal).

At first, the government activated this new educational establishment as a separate school within the Moscow State University, with the foundational decree of the USSR Council of Ministers (No. 2538, November 25, 1946) signed by Stalin.[9] The arrangement did not work smoothly. Consequently, on September 17, 1951, new Ordinance No. 3517-1635 of the Council of Ministers reorganized Fiztekh into an independent Moscow Physical-Technical Institute.[10]

In the early 1970s, Fiztekh had six schools, or faculties (*fakul'tet*), focused on different areas of physics. Each fakul'tet enrolled about one hundred new students, the freshmen, every year. Females constituted on average one-tenth of the students.

From the early days, admission to the institute required excelling in particularly difficult and competitive entrance exams in physics (written and oral) and mathematics (written and oral). For many years, the government allowed the pupils of the secondary schools graduating with distinction (known as the "gold medal" and "silver medal") to enroll into the institutions of higher learning without entrance exams. Fiztekh was an exception and did require entrance exams for everybody.

At each exam, an applicant earned a grade and had to achieve a cer-

---

8. Khristianovich, 1996
9. Karlov, 2006, pp. 16-19
10. Karlov, 2006, pp. 112, 113

tain overall score to be admitted. Some passed (that is did not fail at) the entrance exams at Fiztekh but did not earn sufficient grades for admission to the institute. Per government-approved rules and from MFTI's early days, they "had the right to enroll without further entrance examination to mechanical-mathematical and physics faculties of universities and engineering educational institutions."[11] This rule was in place throughout the entire history of Fiztekh during the Soviet times. MFTI students who were dismissed during their studies for poor performance were usually transferred to and welcomed by numerous other leading engineering institutes of higher learning.

Several instructional, laboratory, and administrative buildings formed the MFTI campus (Fig. 1.6). The nearby dormitories provided basic accommodations for all students, which was most unusual in the country, with shortages of everything. The overwhelming majority of students lived in them during the first two or three years of studies. Lectures and other classes typically ran from 9 a.m. until 5 p.m. every day and sometimes until 6:30 p.m. Students did homework and self-studied later in the evening. Other institutions of higher learning had fewer hours of instruction. In Fiztekh, students also studied for the full six years rather than the typical five or five and a half years.

MFTI assigned third-year students, juniors in the American terminology, to groups linked to various research and development organizations. Such an external center, institute, or design bureau was called a *base institution* or simply a *base* (*baza* in Russian). The third-year students spent one day each week at their base, attending lectures by leading specialists working there.

The fraction of time at the base organizations gradually increased. During the fifth year of studies, the students spent four full days each week there and already engaged primarily in research under the guidance of science advisors. Informally, students called such an advisor a "shef," which roughly corresponded in meaning to a "chief" or "boss" in English or "el jefe" in Spanish. Only one day each week the fifth-year students went to the main MFTI campus in Dolgoprudny to attend the remaining institute-wide coursework, including mandatory reserve officer military training and Marxist-Leninist indoctrination classes. During the last sixth year, the students spent their entire time at the bases working on their master's theses.

The compulsory national military service required every male subject to serve in the armed forces either as an enlisted man for two or three years or as an officer. Correspondingly, a man could not become a scientist

---

11. Karlov, 2006, pp. 24, 25

1. From Fiztekh to IKI   Mike Gruntman

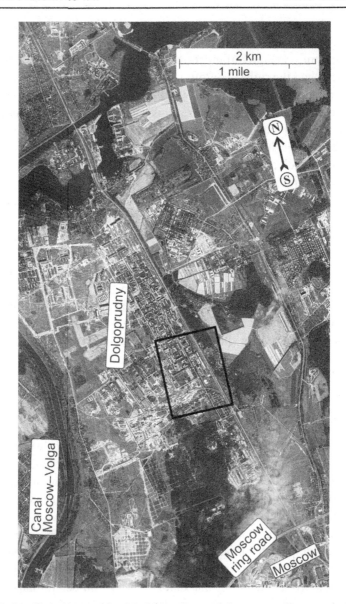

Fig. 1.5. Satellite photograph of Dolgoprudny and the surrounding area in 1976. The Moscow Physical-Technical Institute (black box; magnified in Fig. 1.6) is 2.5 km north from the Moscow administrative boundary, the ring road, and close to the canal connecting the Moscow River with Volga 100 km (60 miles) to the north. Original satellite reconnaissance photograph by KH-9 camera (Mission 1212; September 2, 1976) available from the U.S. Geological Survey; photograph identification, interpretation, and processing by Mike Gruntman.

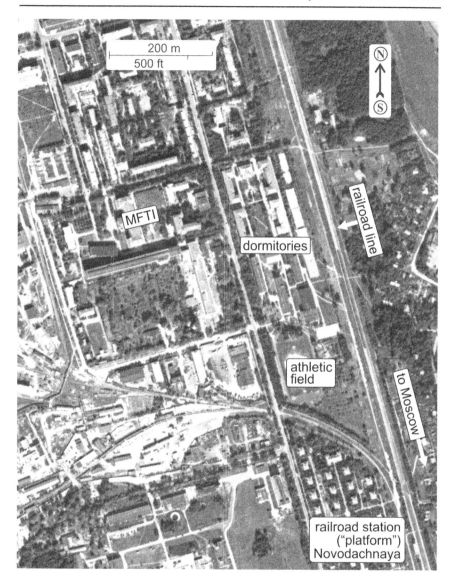

Fig. 1.6. Satellite photograph of the Moscow Physical-Technical Institute (55°55.8' N, 37°31.1' E), MFTI, in 1976. The institute spread over an area next to the railroad station ("platform") Novodachnaya in the town of Dolgoprudny. Suburban trains provided the primary means for reaching Fiztekh. The box in Fig. 1.5 shows the location of this photograph. Original satellite reconnaissance photograph by KH-9 camera (Mission 1212; September 2, 1976) available from the U.S. Geological Survey; photograph identification, interpretation, and processing by Mike Gruntman.

or engineer or medical doctor in the Soviet Union without going through mandatory reserve officer training at the institutions where they studied. Half of the institutes had such military programs.

One could not be idle in a totalitarian country. The government criminally persecuted those who did not work or study after completion of secondary school. Therefore, young men who began to work or tried and failed to gain admission to institutes in competitive entrance exams were conscripted at the age of eighteen. The military also immediately drafted as soldiers the students dismissed from studies in universities and institutes for any reason. Male graduates of institutions of higher learning without reserve officer training, such as actors, teachers, and artists, also served as enlisted men for one year after receiving their degrees.

Military training of students lasted throughout all the years of their studies and culminated with a month or two in a military unit. This latter assignment combined basic military training and familiarization with specialized weapon systems in the field.

Since the early 1960s, the program in Fiztekh led to conferring the rank of reserve officer in the *Strategic Rocket Forces* upon graduation. In the beginning, MFTI received for training on campus a couple of the first Soviet operational ballistic missiles, R-2 (8Zh38),[12] or SS-2 (Sibling) as they were known in the West. In my time, the students of my faculty specialized in servicing liquid-propellant rocket engines of the intermediate-range ballistic missiles (IRBM) 8K63 (R-12), or SS-4 (Sandal).[13]

Yuzhnoe Design Bureau in Dnepropetrovsk (Dnipro, Dnipropetrovsk) in Ukraine developed SS-4 in the late 1950s (Figs. 1.7, 1.8). These nuclear-armed mass-produced missiles used storable propellants and relied on an entirely autonomous guidance system. They became the core of the Soviet IRBM force, with more than 550 deployed by the mid-1960s. In total, the industry built almost 2300 SS-4 missiles.[14] These were the rockets that Soviet leader Nikita S. Khrushchev famously referred to as coming from factory lines as "sausages."

The technology of the SS-4 missiles had become obsolete by the mid-1970s but they remained operationally deployed until the late 1980s. The next cohort of MFTI students coming after my class learned the new generation intercontinental ballistic missiles (ICBM) 8K64 (R-16), or SS-7 (Saddler), also designed and built by Yuzhnoe.

---

12. Shchuka, 2012, p. 392
13. Gruntman, 2004, pp. 286-289; Konyukhov, 2009, p. 51-61; Gruntman, 2015a, pp. 199-201
14. Konyukhov, 2009, p. 58

Fig. 1.7. Left: 22.1-m (72.5-ft) long single-stage liquid-propellant intermediate-range ballistic missile R-12 (8K63, SS-4, Sandal) on a launch stand. The fully-fueled missile weighed 42,000 kg (92,600 lb) and could carry a 1600-kg (3500-lb) nuclear warhead to a range of 2000 km (1250 miles). Photograph courtesy of Yuzhnoe Design Bureau (Office), Dnepropetrovsk, Ukraine.

Fig. 1.8. Above: SS-4's tail section with four nozzles and external control rudders (top right), engine's combustion chamber and nozzle (left), and turbopump assembly (bottom right) at the Strategic Rocket Forces museum in Pervomaisk, Ukraine in 2017. The engine used storable non-cryogenic components: kerosene as a fuel and nitric acid with the addition of nitrogen tetroxide as an oxidizer. Photographs courtesy of Mike Gruntman.

1. From Fiztekh to IKI — Mike Gruntman

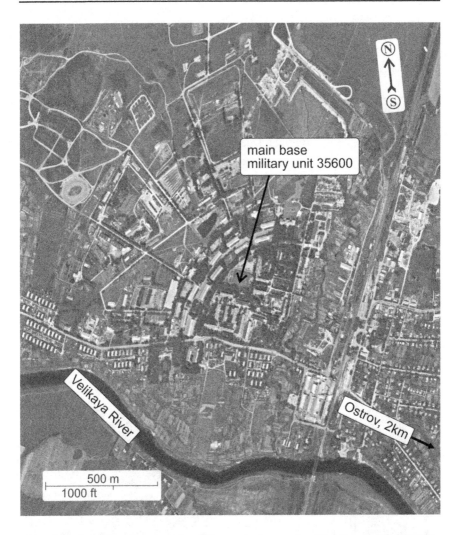

Fig. 1.9. Satellite photograph of the main base of military unit No. 35600 (57°21.5' N, 28°18.4' E) of the Strategic Rocket Forces near the town of Ostrov 600 km (370 miles) west-northwest from Moscow. The base conducted reserve officer training of students of Fiztekh and some other institutions of higher learning. A field position, *Beryoza* (57°24.0' N, 28°14.5' E), with the intermediate-range ballistic missiles SS-4 was 4 km (2.5 miles) away. The unit also supported three launch positions of operational ballistic missiles farther afield. Original satellite reconnaissance photograph by KH-9 camera (Mission 1212; July 24, 1976) available from the U.S. Geological Survey; photograph identification, interpretation, and processing by Mike Gruntman.

After the fifth year at Fiztekh, we spent one month of training in military unit No. 35600 of the Strategic Rocket Forces near the town of Ostrov in the Pskov region, 600 km (370 miles) west-northwest from Moscow and 30 km (20 miles) from the border of modern-day Latvia. This IRBM unit included the main base (Fig. 1.9), one "fixed field site" for training, and three operational missile launch areas, 10, 18, and 29 km away.[15]

Our "military month" began with close-order drills, marching in platoon formations, "kitchen patrols," and firearms (Fig. 1.10). Then, we moved to a fixed field site, *Beryoza* (birch tree), 4 km (2.5 miles) to the northwest for training with the ballistic missiles. Interestingly, our studies never included any details of the warhead. I saw this part of the SS-4 without a nuclear charge for the first time in a museum only many years later (Fig. 1.11).

The government considered Fiztekh graduates so valuable for the country's science and defense establishments that it exempted all from being called up for regular military service. Many graduates of other universities and engineering institutions served as junior officers for two years after completing their studies.

The Communist Party mandated annual Marxist-Leninist indoctrination coursework for students in all universities and institutes across the country. Incidentally, achieving Ph.D. degrees also required an extra one-year course and an unavoidable exam in Marxist-Leninist philosophy, in addition to screening exams in a chosen science field and a foreign language.

The communist country compelled MFTI students to do manual physical labor on construction projects for a month during one summer. Soviet propaganda habitually denigrated bourgeois "rotten intel-

Fig. 1.10. The author of this book during mandatory one-month military training in a unit of the Strategic Rocket Forces in the summer of 1976. Photograph from collection of Mike Gruntman.

---
15. CIA, 1965, 1969

1. From Fiztekh to IKI   Mike Gruntman

Fig. 1.11. The author of this book for the first time "meets" an SS-4 warhead in Swedish Military Museum in Stockholm in 2016. Photograph from collection of Mike Gruntman.

ligentsia" and praised manual labor as ostensibly strengthening ideological bearings of the youth. Such an approach resembled on a smaller scale the practice of *Reichsarbeitsdienst*, or the *Reich Labor Service*, in National-Socialist Germany in the 1930s, another sister regime anchored in a radical socialist ideology.

Other European communist countries also followed such practices to various degrees. In the late 1950s, Hungary, for example, introduced the requirement for students to work 20 hours each year on construction sites.[16] Fortunately, the early ideological vigor and fervor had largely dissipated by my time, giving place to corrupt atrophy in such endeavors. Not in the People's Republic of China, however, where the ongoing enthusiastic wave of atrocities of the "Great Proletarian Cultural Revolution" has been devastating the country.[17]

Fiztekh organized compulsory manual work, the so-called "constructions detachments," for one month during the first summer vacation after the freshman year (Fig. 1.12). The experience had some useful consequences. In a common practice during vacations in the following summers, many students voluntarily formed such groups and engaged in construction projects in remote parts of the country to earn money.

My faculty, FAKI (Fig. 1.13), assigned us to base organizations at the end of the second year of studies in the spring of 1973. Design bureaus and research and development institutes of the ballistic missile and space industry had a "controlling interest" of the faculty and dominated the FAKI

16. e.g., Golovanov, 2001, v. 1, p. 111
17. Courtois et al., 1999, pp. 513-538; Gruntman, 2004, pp. 442-444

Fig. 1.12. The author of this book at the mandatory manual physical work in a construction detachment during the summer of 1972, part of the education of physicists. From collection of Mike Gruntman.

Fig. 1.13. Old emblems of the Moscow Physical-Technical Institute MFTI (top) and Faculty of Aerophysics and Space Research FAKI (bottom).

bases. A couple of institutes, however, belonged to the Academy of Sciences, including the Space Research Institute.

Fiztekh included me, as I asked, in a group heading to IKI that usually consisted of 5-7 students each year. The new fall semester of 1973 would be our first in the mysterious base organization.

One day, a secretary of the FAKI dean "volunteered" me to deliver to IKI some administrative papers. "They have changed the director of the institute," she said and mangled the name of the just-appointed director Roald Sagdeev, a rising to prominence 40-year-old academician. This coveted title of an elected full member of the USSR Academy of Sciences required a combination of scientific accomplishments and some political loyalty, especially for serving as an influential institute director.[18]

So I came to the Space Research Institute at the same time as Sagdeev who would transform the institute and lead it until 1988.

---

18. Gruntman, 2015a, p. 87

New institute in the Academy of Sciences

The Academy of Sciences had opened the new Space Research Institute several years before the described events. The institute formed its staff by drawing scientists from other research organizations. IKI rapidly grew and evolved administratively. The Academy of Sciences appointed a prominent specialist in gas dynamics, academician Georgii I. Petrov as the first institute director (Fig. 1.14).

On May 4, 1965, Chairman of the USSR Council of Ministers Aleksei N. Kosygin sent a draft of the government decree approving the proposal by the Academy of Sciences to establish IKI to the highest body of the country, the Central Committee of the Communist Party of the Soviet Union. The Presidium of the Central Committee approved the draft on May 15, 1965, and the Council of Ministers issued the top-secret decree No. 392-147, signed by Kosygin, on the same day.[19]

The government gave two months to the Academy of Sciences to work out, in coordination with the *Ministry of General Machine Building* and *Ministry of Defense*, the "Statute of the Space Research Institute of the Academy of Sciences," and then the overseeing powerful *Military-Industrial Commission*[20] to approve it. The Ministry

Fig. 1.14. First four directors of IKI: Georgii I. Petrov, 1965-1973 (top left); Roald Z. Sagdeev, 1973-1988, (top right); Al'bert A. Galeev, 1988-2001 (bottom left); and Lev M. Zelenyi, 2002-2017 (bottom right). Anatoly Petrukovich has directed IKI since 2018. Photographs from Zelenyi, 2015, p. 27.

---

19. Baturin et al., 2008, pp. 298-299
20. Gruntman, 2015a, p. 96

of General Machine-Building, known by its Russian abbreviation MOM, has directed work on ballistic missiles, rocketry, and space in the country since the mid-1950s.

In the top-down central-planning bureaucratese, the decree of the Council of Ministers authorized, "as an exception," to begin construction of the institute buildings immediately:

> 5. Authorize the USSR Academy of Sciences to build in 1965-1967, as an exception, laboratory buildings with the working area up to 30,000 square meters [320,000 sq. ft.] for the Space Research Institute in Moscow.
>
> [Order] the municipal executive council of Moscow to allocate land to the USSR Academy of Sciences for construction of the mentioned buildings.
>
> Assign construction of the buildings of the Space Research Institute to the Glavspetsstroi organization of the State Committee for Special Assembly and Construction Works of the USSR.
>
> 6. Authorize the USSR Academy of Sciences and the USSR State Planning Committee to include, as an exception, the construction of laboratory-production buildings of the Space Research Institute into the construction plan for 1965-1966 without properly approved design and cost estimate documentation...
>
> [Order] the State Bank of the USSR to provide financing of the construction of the mentioned buildings according to design and financial estimates, based on working blueprints until the design and cost estimate documentation is [properly] approved...[21]

Soviet managers made many ad hoc operational decisions that reflected the power structure and access to the individual ruling top members of the Communist Party and government. Consequently, economic plans and their implementation had to be continuously adjusted. Such deviations from plans, manifested by the "as an exception" provisions in the decree to establish IKI, permeated numerous ordinances, findings, and decisions on all levels of the ostensibly strictly planned state and its various organizations. This inevitable feature of socialism resulted from collisions of ideological and bureaucratic fantasies with the realities of life. History should have provided a lesson, and perhaps given pause, to the growing number of central-planning fans in the free world. They remain, however, enthusiastically inspired by the dysfunctional Marxist-Leninist-Maoist utopia.

21. Baturin et al., 2008, pp. 298-299

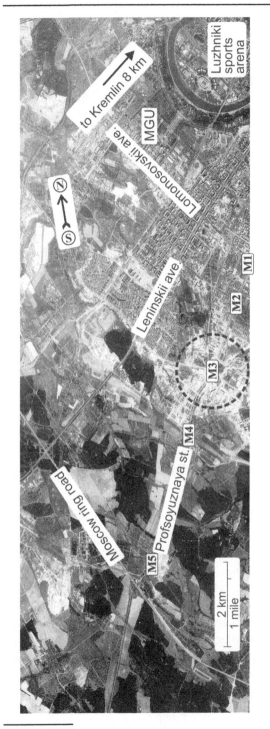

Fig. 1.15. Satellite photograph of the south-southwest area of Moscow in 1966. Rectangles with the letters "M" show locations of existing at the time and future metro (subway) stations: M1 – Profsoyuznaya; M2 – Novye Cheremushki (Cheryomushki); M3 – Kaluzhskaya; M4 – Belyaevo; M5 – Teplyi Stan). The dashed circle shows the area at the intersection of Profsoyuznaya and Obrucheva streets (see Fig. 1.16) chosen for IKI. The Moscow ring road, approximately 34 km (21 miles) in diameter (Fig. 1.3), formed the administrative boundary of Moscow. The MGU label shows the main site of Moscow State University, abbreviated MGU in Russian, across the Moscow River from the Luzhniki sports arena. Original satellite reconnaissance photograph by KH-4 camera (Mission 4030; July 16, 1966) available from the U.S. Geological Survey; photograph identification, interpretation, and processing by Mike Gruntman.

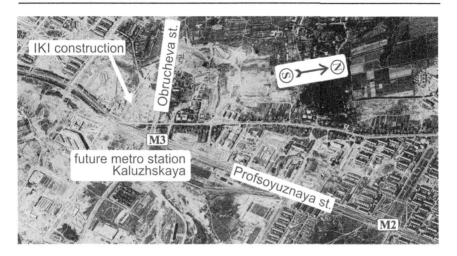

Fig. 1.16. Area of the construction site of the main IKI building (55°39.3' N, 37°32.0' E) at the intersection of Profsoyuznaya and Obrucheva streets (dashed circle in Fig. 1.15) near the future metro station Kaluzhskaya (M3) in 1966. M2 – Novye Cheremushki metro station. Original satellite reconnaissance photograph by KH-4 camera (Mission 4030; July 16, 1966) available from the U.S. Geological Survey; photograph identification, interpretation, and processing by Mike Gruntman.

Construction of the institute buildings quickly began. The Moscow geography at the time substantially differed from the present day. The selected site at the crossing of Profsoyuznaya and Obrucheva streets was at the city's outskirts. Figures 1.15 and 1.16 show satellite photographs of that part of south-southwest Moscow in July 1966. The construction site occupied the southwest corner of the street intersection (Fig. 1.16). At that time apartment buildings reached only the present-day subway (Moscow metro) station "Belyaevo." Agricultural fields stretched beyond that area to the ring road (Fig. 1.15) that defined the administrative boundary of the city. Today, residential areas fill all space to the ring road and beyond.

The construction progressed and another satellite photograph (Fig. 1.17) four years later in July 1970 clearly shows the long main building of the institute as well as small two-story *steklyashka* ("glassy") houses adjacent to its backyard. Figure 1.18 shows the view of that interior side of IKI with these glassy houses in front and the prominently extending supporting parts (elevators, piping, communications) of the main building. Other structures that house the institute's engineering services spread father to the left along all its length.

1. From Fiztekh to IKI   Mike Gruntman

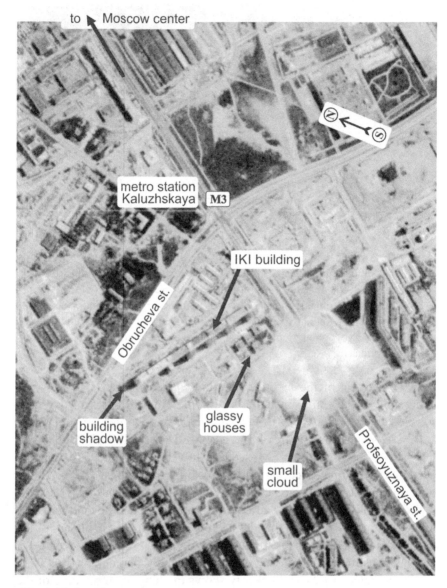

Fig. 1.17. Satellite photograph of the 1300-ft (400-m) long IKI building erected at the intersection of Profsoyuznaya and Obrucheva streets in 1970. Small two-story steklyashka ("glassy") houses (Fig. 1.18) are adjacent to the backyard of the building. M3 is the metro station Kaluzhskaya. Original satellite reconnaissance photograph by KH-4 camera (Mission 1111-1; July 26, 1970) available from the U.S. Geological Survey; photograph identification, interpretation, and processing by Mike Gruntman.

Fig. 1.18. Backyard side of the main IKI building with the *steklyashka* "glassy houses" (55°39.2' N, 37°32.1' E) in the 1970s or 1980s. Frame from a documentary film, *Honeymoon on the Moon*, CineGiraffe, 2016, https://youtu.be/7wcTfhjSNyw; accessed on April 20, 2021.

In another pathology of "scientifically organized" centrally planned society, construction of the main building got mired in bureaucratic crosscurrents, and completion was not assured, despite being authorized by the highest government bodies. Institute director academician Petrov had to bring in a powerful military organization to occupy the eastern section of the 1300-ft (400-m) long building where its side faced Profsoyuznaya street (Fig. 1.19). The military provided additional political and administrative clout and resources that helped complete the construction. This influential neighbor evolved into the main procuring arm of the Soviet, and now Russian, space forces known as GUKOS (*Glavnoe Upravlenie Kosmicheskikh Sredstv*, or *Chief Directorate of Space Assets* of the Ministry of Defense).

The IKI part of the building had been fully operational by 1973 when I walked in through the institute doors for the first time (Fig. 1.20). Seven spacious high-ceiling working floors housed offices and laboratories, with the so-called technical floors with somewhat lower ceilings and smaller windows in between (Fig. 1.19).

The construction of IKI, in addition and "as an exception" to the national five-year plan, relied on an existing design of an unrelated research building optimized for the chemical industry.[22] Its technical floors were

---

22. Shalimov, 2006, p. 27

Fig. 1.19. Main IKI building 400-m (1300-ft) long as viewed from across Profsoyuznaya street in June 2016. Rows of large and narrow windows correspond to high-ceiling working floors (offices, laboratories) and low-ceiling supporting "technical" floors, respectively. The leftmost (eastern) section houses a military organization, GUKOS, responsible for procuring space assets for the Ministry of Defense. Photograph by Mike Gruntman.

Fig. 1.20. The author of this book at the main entrance of IKI in 2018, forty-five years after he walked through these doors for the first time. From collection of Mike Gruntman.

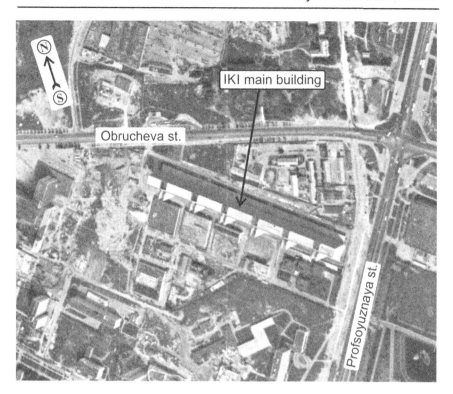

Fig. 1.21. Satellite photograph of IKI in 1977. The main 400-m (1300-ft) long building cast a shadow to the north-northwest. Therefore, the photograph was taken about one hour before the local noon. One can see several supporting structural parts at the back of the building (Fig. 1.18). Original satellite reconnaissance photograph by KH-9 camera (Mission 1213-1; July 22, 1977) available from the U.S. Geological Survey; photograph identification, interpretation, and processing by Mike Gruntman.

intended to supply various reactive liquids and gases to the working floors. IKI did not need such engineering infrastructure. Therefore, the technical floors initially remained empty and were occasionally used for minor storage. The institute later remodeled them into offices.

Multiple smaller structures spread over open territory at the back of the main institute building (Figs. 1.21). They housed computer facilities, test vacuum chambers, utilities, and other supporting engineering services. With time the administration fenced the area off, forming the backyard of the institute with controlled access.

A CIA report[23] described in 1980 that "IKI consists of 21 buildings with a total floorspace of 131,177 square meters." It estimated the main research building to be 403 m long, 18 m wide, and 52 m high with 116,064 square meters of floorspace. The document also noted that while the institute was focused on open space science research, a "classified research program was begun in 1977 ... to develop large space-based antennas to monitor US microwave communications and radars." The institute indeed restricted access to one section of the fifth working floor of the building for classified research, with an additional guard at the entrance door leading to that area.

Students at IKI

Our student life in the Space Research Institute began with specialized courses taught by IKI staff scientists. Research under the guidance of science advisors followed. The courses gave us unique opportunities to meet the country's leading space specialists and learn about some of the institute's programs. Several instructors were deputy directors and heads of research departments and laboratories. Knowing these accomplished and influential specialists personally made access to them easier in the future. Teaching responsibilities did not particularly burden IKI scientists and provided them with excellent additional earnings.

Administratively, MFTI set up a *kafedra* of space physics at IKI. In the Soviet Union, a kafedra (from the Greek καθέδρα, or a *seat*) was equivalent in scope to a specialty within an academic department of an American university but often with a larger number of instructors. It had its own chairperson and faculty. While the space physics kafedra formally belonged to MFTI, its faculty members worked full-time as staff members at IKI. They taught specialized courses typically once a year and oversaw student research in addition to their "day jobs." The arrangement resembled in some respect adjunct faculty in U.S. universities.

Incidentally, the concept of tenure for faculty in universities and institutes did not and could not exist in communist countries. Academic freedom and consistently implemented socialism are fundamentally incompatible. Ironically, this rude awakening will welcome the present-day proponents of socialism in American universities if they ever succeed politically. It will be too late for them, though.

Many leading IKI specialists lectured to us, including Al'bert A. Galeev, Vitalii D. Shapiro, Valentin S. Etkin, and IKI deputy directors Yulii K. Khodarev and Valerii G. Zolotukhin. Galeev and Shapiro joined the insti-

---

23. CIA, 1980

tute shortly after Sagdeev became director in 1973. Galeev would succeed Sagdeev and serve as the institute director from 1988-2001 (Fig. 1.14). Another future IKI director (2001-2017) and vice-president (2013-2017) of the Russian Academy of Sciences, Lev Zelenyi was a Ph.D. student in the mid-1970s. He soon became a research staff scientist and would later lecture to students.

Figure 1.22 shows the final exam in the course "Space Electrodynamics" taught by Al'bert Galeev to our group in 1975. The instructor's title and signature after the problems look odd, reflecting the bureaucratic culture of the country.

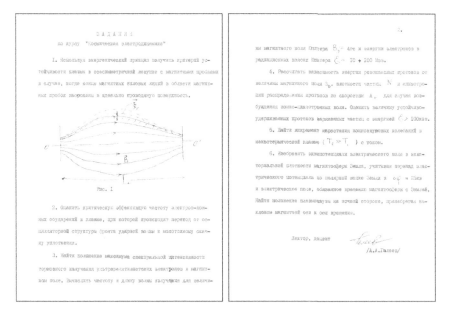

Fig. 1.22. Typewritten final exam problems in the course "Space Electrodynamics" taught by Al'bert Galeev in the fall semester of 1975. The characters are blurred as typical for a carbon copy. From collection of Mike Gruntman.

Institute director Sagdeev also taught one course, but mainly through his associates. Obviously, he was a very busy man. With time, the institute would grow to two thousand employees, including those in a few branches located elsewhere in the country. We saw Sagdeev only twice during that semester: at the very first lecture and at the last one when he distributed problems for the final exam.

1. From Fiztekh to IKI   Mike Gruntman

IKI engaged in highly visible and expensive space programs. Such activities understandably required continuous interactions with powerful government officials and thus the undivided attention of Sagdeev. It always puzzled me that some senior scientists could not sincerely comprehend the scope of his administrative responsibilities and the associated effort. Interestingly, many of my present-day faculty colleagues also show a similar lack of understanding and appreciation of the demanding work and skills of managers and administrators. Such carping seems to be a common trait of intellectuals venturing beyond their domains of expertise—the people who never experienced responsibilities of significant executive power.

Fig. 1.23. Forty-six years after the first meeting in 1973: Roald Sagdeev (right) and the author of this book (left) at the International Astronautical Congress in Washington, D.C., in October 2019. Photograph from collection of Mike Gruntman.

While I have never worked directly for or with Roald Sagdeev, he has known me since those student years. Sagdeev always returned a friendly greeting when he occasionally saw me in the halls of the institute. Several years later I defended my Ph.D. thesis at his science council.

The rigidly hierarchical Soviet system determined who could see whom and when. Since I have never been directly involved in his "personal" projects, I entered Sagdeev's office less than ten times during all my years in the institute. Even these rare meetings with the director were more frequent than those experienced by many of my peers in the same age group. Sagdeev knew about my work to a large degree from my foreign collaborators.

Many years later, our paths crossed on the other side of the globe (Fig. 1.23). As my interests expanded to spacecraft technologies, I helped Sagdeev in the mid-1990s to connect to some specialists in the U.S. space industry. Several years later, we served for one year on a government committee evaluating a somewhat controversial research and development NASA program. I also once paid a friendly visit to him at the University of Maryland during one of my business trips to the Washington, D.C., area. The interactions were always courteous and friendly.

In 1976, MFTI expanded by adding a new *Faculty of Problems of Physics and Energetics* (*Fakul'tet Problem Fiziki i Energetiki*, or FPFE). This new faculty administratively took over the kafedra of space physics and IKI as one of its base institutes. A celebratory ceremony of a formal opening of FPFE took place in a "glassy house" (Fig. 1.18) next to the IKI backyard in February 1977. Attending dignitaries (Fig. 1.24) included two Nobel prize winners in physics, Aleksandr (Alexander) Prokhorov (1964) and Petr Kapitsa (he would receive his Nobel prize the next year for the discovery of superfluidity of liquid helium in the late 1930s).

We graduated in a mere few months. My group became the last cohort of MFTI students at the Space Research Institute with degrees from FAKI. Henceforth, Fiztekh students studying at IKI would graduate from the FPFE faculty.

Science advisor or *shef*

IKI assigned each student to his or her science advisor, the shef, after one year of studies in the institute. During our first year, Leonid L. Van'yan managed the space physics kafedra. He likely served as the deputy chairman because in a semi-feudal system, the institute director had to also be, *ex officio*, the kafedra chairman. Van'yan talked to me about my interests and gave a few articles from the *Journal of Geophysical Research*, JGR,

Fig. 1.24. Formal opening of MFTI's new Faculty of Problem of Physics and Energetics, FPFE, in a glassy two-story building adjacent to the IKI backyard in February 1977. FPFE took over from FAKI the Space Research Institute as a base organization and its kafedra of space physics. From 1978 to 2016, Fiztekh students at IKI would graduate from FPFE. In a major administrative reorganization, MFTI rearranged its faculties into several schools in the late 2010s.

Front row, left to right:
- IKI Director Academician Roald Z. Sagdeev (with eyeglasses);
- Academician Aleksandr M. Prokhorov (Nobel Prize in physics, 1964);
- FPFE Dean Academician Evgeny M. Velikhov (a prominent leader of Soviet weapons programs and head of a branch of the Kurchatov Institute of Atomic Energy in Troitsk near Moscow; vice-president of the USSR Academy of Science from 1978-1996);
- Academician Ivan F. Obraztsov (Higher Education Minister of the Russian Federation from 1972-1990);
- Academician Petr L. Kapitsa (collaborator of Ernest Rutherford in the 1920s and 1930s and one of the most influential founders of MFTI; Nobel Prize in physics, 1978);
- MFTI Rector (President) Academician Oleg M. Belotserkovsky.

The author of this book, Mike (Misha) Gruntman, with a mustache, tie, unbuttoned jacket, and his right hand on his hip, is in the back row seen between Sagdeev and Prokhorov.

Photograph from collection of Mike Gruntman.

to read during my summer vacations. He planned to engage me in research under his supervision in the next fall semester.

When we got back to IKI in early September 1974, Van'yan "disappeared." Administrative reorganizations by the new forceful director Sagdeev gradually propagated through the institute and shipped my tentative future shef away to some other organization. I never saw or talked to him again.

Fortunately, the students were insulated from institute politics. At first, we had no clue about frictions and struggles among many bright personalities and accomplished scientists who populated IKI. With time, this knowledge "gap" would be quickly filled.

Vladas Leonas taught one of our very first courses at IKI. An expert in atomic physics, he focused on experimental techniques in space research. His broad review-type lectures on measurements in space particularly appealed to me because of my one peculiar personal habit.

From the very first month as a Fiztekh student, I periodically visited the institute library in Dolgoprudny and perused and read articles in the new issues of the leading Soviet physics review journal *Uspekhi Fizicheskikh Nauk* (*Successes of*, or *Advances in*, *Physical Sciences*). Since 1958 a U.K.-based learned society, the *Institute of Physics*, IOP, published its cover-to-cover English translation as *Soviet Physics Uspekhi* (*Physics-Uspekhi* since 1993). This high-quality journal is similar to *Reviews of Modern Physics* published by the *American Physical Society*.

Much of the journal content extended beyond student understanding, but the rigorous articles by top physicists provided a window into a fascinating world of physics. The journal triggered curiosity and interest in various areas, which would influence my work in the future.

Leonas and I established very good personal relations. He was an intelligent, well-educated, and decent man with broad interests and a sense of humor. It also became clear that he did not like the Soviet regime. These circumstances determined, to a significant degree, the beginning of my professional life.

Vladas Leonas became my science advisor in September 1974. I worked with him for the next 15 years.

2. Department of Space Gas Dynamics

Semi-independent fiefdom

The experimental laboratory of Vladas B. Leonas constituted half of the Department of Space Gas Dynamics, or Department No. 18, of IKI. Vladimir B. Baranov oversaw the other, theoretical laboratory. Academician Georgii I. Petrov presided over the department, with its two laboratories counting 11-14 staff members each.

A highly accomplished specialist in gas dynamics, Petrov served as the first, founding director of IKI from 1965–1973 (Fig. 2.1). As a full member of the USSR Academy of Sciences, he had the privilege to have a research department in one of the institutes of the academy. When Sagdeev replaced him as IKI director in 1973, Petrov retained Department No. 18 as "his own." Only three or four academicians worked in the institute at any given time. The exalted status in the science establishment allowed them to maintain semi-independent fiefdoms that the institute director could not fully control.

Petrov usually appeared at IKI once a week and chaired a science seminar of the department in the afternoon. Baranov's theoreticians focused on gas dynamics and its applications to space research, which was among the science interests of Petrov. In contrast, the experimental laboratory of Leonas concentrated on the physics of atomic collisions and atomic and molecular interactions, areas outside the expertise of Petrov and Baranov.

Theoretical gas dynamics dominated the seminar agenda. Experimentalists rarely attended these meetings, except for laboratory head Leonas. The department's experimental and theoretical groups occupied offices on different floors of the large IKI building and did not interact much, as their science interests had little in common.

2. Department of Space Gas Dynamics — Mike Gruntman

Fig. 2.1. First (founding) director of IKI, 1965-1973, and head of the Department of Space Gas Dynamics, or Department No. 18, aerodynamicist Academician Georgii I. Petrov in May 1982. Photograph from collection of Mike Gruntman.

There were occasional exceptions. When theoreticians needed a screwdriver or a hammer to repair a chair or desk, they would descend from their offices on the seventh floor to borrow the tools. In addition, the institute periodically sent its scientists and engineers for a few days to vegetable storage depots to sort out potatoes or load cabbage. Other similar scholarly assignments included a week at an agricultural farm digging out and picking fodder beets. On such occasions, the administration grouped us and camaraderie developed.

A highly inefficient and wasteful centrally controlled economy inherently suppressed incentives for excellence. Poorly paid people performed accordingly. Some saw the causation in reverse. In any case, unpleasant and unrewarding occupations always needed workers.

Socialist mismanagement and social engineering particularly devastated agriculture and related production areas. Therefore, the regional Communist Party bodies assigned quotas to the Academy of Sciences, industry, and institutions of higher learning to periodically send their employees and students to sort out already rotten vegetables in storage depots in the cities and to work on agricultural farms in rural areas. Or they would clean debris and carry bricks at construction sites. The Party had imposed such duties on the population since at least the 1950s.[1] In Central Asia, educated inhabitants of cities, townsfolk, students, and secondary school pupils spent many weeks picking cotton. These practices especially hurt the most productive professionals.

Heads of laboratories Baranov and Leonas (Fig. 2.2) acted rather independently at IKI, enjoying some protection inside Department No. 18, the fiefdom of their "own" Academician Petrov. This inevitably led to mount-

---

1. e.g., Golovanov, 2001, v. 1, p. 59; v. 2, p. 358; v. 3, p. 11

Fig. 2.2. Heads of theoretical and experimental laboratories in Department No. 18 of IKI Vladimir Baranov (left) and Vladas Leonas (right), respectively. In the background next to Leonas are Hans Fahr (University of Bonn) in the center and Darrell Judge (University of Southern California). Photographs (October 1988) courtesy of Mike Gruntman.

ing frictions with the new forceful and energetic institute director Sagdeev. Finally, in 1987, as told in Chapter 11, the Academy of Sciences transferred the entire department of Petrov to the quieter environs of the *Institute for Problems in Mechanics*, or IPM.

Initially, internal IKI politics only marginally affected me, a junior researcher. Administrative struggles did not look important as I concentrated on work. It became clear later, however, that friendlier relations with the institute director and more business-oriented and "politically" savvier laboratory heads would have facilitated the development of instrumentation and space experiments that I advocated and worked on. The advances in position-sensitive detectors and energetic neutral atoms could have been much faster. These boundary conditions, or constraints, were obviously beyond my control and pay grade. As they say, we do not choose our parents.

2. Department of Space Gas Dynamics   Mike Gruntman

Atomic collisions laboratory

The laboratory of Leonas consisted of two parts. I soon attached myself to a group studying collisions and differential scattering of neutral atoms with "high" energies (Fig. 2.3). In our context, "high" meant "non-thermal" energies between a few hundred and a few thousand electronvolts, or eV.[2]

In this area of atomic physics, the studies of particle collisions probed the potentials of interatomic and intermolecular interactions. Measurements of differential scattering of colliding atoms with energies 600–2000 eV on small, milliradian[3] angles opened ways to determine interatomic interaction potentials in the range from a fraction of one to several eV, corresponding to interparticle distances from one half to a few angstroms.[4] Many physics applications relied on knowledge of interaction potentials, including calculations of transport coefficients and other properties of gases and plasmas under nonequilibrium conditions.

The high-energy group included Aleksandr (Sasha) Kalinin (Fig. 2.3) and Vladimir (Volodya) Khromov. It occupied a room on the fourth floor of the IKI building next to the office of Leonas. The laboratories headed by Oleg L. Vaisberg, Yuri I. Galperin, and Lev M. Mukhin and their offices were a few steps away. A group of Georgii G. Managadze had laboratories and offices farther away on the same floor in a different section of the building.

My new neighbor Vaisberg would later become an official reviewer of my master's thesis. Future IKI director Al'bert A. Galeev also occupied an office nearby. These neighbors, instructors in our classes, and several other IKI specialists with whom I interacted, such as Konstantin I. Gringauz, Vladimir G. Kurt, and Rashid A. Sunyaev, truly represented the who-is-who of Soviet civilian space science at that time.

The other part of the laboratory of Leonas experimented with beams of "low-energy" atoms and molecules, the so-called "thermal beams" with energies less than 1 eV. Evgenii N. Evlanov led the low-energy beam group (Fig. 2.4). One of its main challenging programs aimed to measure kinetic energies and vibrational levels of molecular hydrogen leaving cold surfaces after formation in the recombination of atomic hydrogen. Such processes on surfaces of interstellar dust grains could play an important role in determining properties of the interstellar medium and presented a fundamental interest for astrophysics.

---

2. One electronvolt is energy gained by an electron accelerated by a voltage difference of 1 volt; $1 \text{ eV} = 1.602 \times 10^{-12}$ erg $= 1.602 \times 10^{-19}$ J.
3. One milliradian is about one-seventeenth of one degree, 1 mrad $\approx 0.057°$.
4. One angstrom is one-tenth of one nanometer, 1 Å $= 0.1$ nm $= 10^{-8}$ cm. The "size" of a hydrogen atom is approximately one angstrom.

Fig. 2.3. Laboratory head Vladas Leonas (leftmost) and the "high-energy" atomic beam group in experimental room No. 419 on the fourth floor of the IKI building in the early 1980s. Next to Leonas are, left to right, Victor Morozov, Alexander Kalinin, and the author of this book, Mike, or Misha in "the previous life," Gruntman. In the background, one can see high-voltage power supplies, electronic pulse counters, the ion beam control panel of a modified mass-spectrometer, and power and signal cables stretched above. Photograph from collection of Mike Gruntman.

Evlanov's group included several scientists, technicians, and engineers, including Sergei V. Umansky (Fig. 2.4), who became a good friend. They occupied a large room on the ground floor of the IKI building, sandwiched between the experimental hall of a group directed by Igor M. Podgorny and a door, with a guard on duty, leading to the IKI backyard. A specialist in electronics, Boris (Borya) Zubkov (Fig. 2.4) supported both groups of the laboratory of Leonas. Later, two MFTI students, Victor (Vitya) Morozov (Fig. 2.3) and Alla Kozochkina, née Demchenkova, and a couple of others joined the laboratory as staff members.

2. Department of Space Gas Dynamics   Mike Gruntman

Fig. 2.4. Head of the "low-energy" beam group Evgenii Evlanov (left, in 1990s); engineer Sergei Umansky (center, in 1980s); and electronics engineer Boris Zubkov (right, in 1990s). Photographs courtesy of Mike Gruntman.

In the early 1980s, the group of Evlanov together with Zubkov and Khromov took part in work on a dust particle impact analyzer, *PUMA* (Fig. 2.5), for the space mission *Vega* to comet Halley. This mission would become one of the highest achievements of the Soviet space science program.

Scientists from the *Max Planck Institute for Nuclear Physics* in Heidelberg, West Germany, led this joint development. They also built a conceptually similar instrument for the European *Giotto* spacecraft to the comet. Work on PUMA led to increased frictions between Leonas and Sagdeev (Chapter 9). Evlanov, Zubkov, and Khromov formed a semi-independent group within the laboratory. They split and stayed at IKI when the department transferred to IPM in 1987.

Rudderless without a helmsman

The Department of Space Gas Dynamics stood out at IKI in one peculiar way: it did not have a single member of the Communist Party. This made the composition of the department's staff most unusual, as advancement in any profession and obtaining various benefits in all aspects of life required Party membership. A totalitarian system demands not only conformity and obedience by the subjects but also their participation in its functioning and acts (and crimes, at times), willing or not, sincere or fake.

Fig. 2.5. Engineering (development) unit of the Soviet-German dust particle impact analyzer PUMA of the Vega mission to comet Halley on display in the IKI museum. The prominent reflectron part of this time-of-flight mass spectrometer is on the left.

The top-left inset shows a cropped-out portion of a group photograph of IKI specialists during the award ceremony in 1986 for the successful Vega mission. There, from left to right in the top row, are Boris Zubkov and Evgenii Evlanov from Department No. 18 and Vadim Angarov from IKI's design bureau branch in Frunze, today's Bishkek, Kyrgyzstan. At the bottom left is Boris N. Yeltsin, the future President of Russia, who then headed the governing Communist Party body of the city of Moscow. Yeltsin presented the awards.

Photographs (2018) courtesy of Mike Gruntman.

Approximately 7-8% of the country's total population, or about 15% of the working-age adults, formally belonged to the Party, the "vanguard of the working class." Article 6 of the USSR Constitution that was adopted in 1977 codified "the leading and guiding role of the Communist Party in Soviet society."[5] In the late 1980s, CPSU counted 19 million members out of 250 million people. Ruling parties of totalitarian socialist regimes, the ideological siblings, show similarities worldwide. For example, the present-day *Chinese Communist Party* (CCP) consists of more than 90 million

---

5. Constitution ..., 1977

2. Department of Space Gas Dynamics   Mike Gruntman

members, which represents a similar fraction of the population of the People's Republic of China. At its peak in the 1940s in another country, the membership of the ruling *National Socialist German Workers' Party*, or NSDAP and commonly referred to as the Nazi Party, reached 10-11% of Germany's population.

Whether one addresses a member of the movement, *Die Bewegung*, as a *comrade, comarada, tovarishch, tong-zhi,* or *Parteigenosse*, they all remain brothers and sisters of the same tight family of radical socialists.

In an exception to the rule, the first IKI director academician Georgii Petrov avoided Party membership. His early scientific contributions of national importance, particularly in supersonic gas dynamics of jet aviation and ballistic bodies entering the atmosphere, allowed him to get away with not being a communist. For others, pressure to conform was relentless.

Both Leonas and Baranov, as well as a few other staff members of Department No. 18, viewed some aspects of the Soviet system critically. One could express these sentiments only among very close and trusted friends. Loose lips could easily sink careers and cost jobs in such a place as IKI. Even more importantly, a politically colored "indiscretion" could close forever any employment in science-related fields.

During my time and in contrast to the earlier days of the Soviet regime,[6] the Marxist-Leninist state had "mellowed" a little bit and stopped physically annihilating millions of not fully conforming people. A human rights activist and prisoner of conscience, physicist Yuri Orlov, described that

> [Soviet s]ociety remained totalitarian, but at least had ceased to wallow in blood and vomit. ... [Soviet leader Nikita S.] Khrushchev reorganized the country [in the late 1950s and early 1960s] from a regime of total self-destruction to a moderately totalitarian one, in which the average citizen could at least die peacefully in his own bed.[7]

The Communist Party kept closed, however, many doors in the professional world to doubters, and it promptly imprisoned selected dissidents, including the author of the quote above. A senior Communist Party official who would later serve as a senior adviser to Gorbachev quoted in his diary a memo by Chairman of the Committee for State Security (KGB[8]) Yuri V. Andropov, who summarized in 1976 that

---

6. e.g., Solzhenitsyn, 1973; Begin, 1977 (1957); Courtois et al., 1999
7. Orlov, 1991, pp. 139, 146
8. *Komitet Gosudarstvennoi Bezopasnosti* (Committee for State Security), or KGB, was the main organization in the Soviet Union responsible for matters of state security and repression.

during the last 10 years, about 1500 people were arrested for anti-Soviet activities. There were 850 political prisoners [in the Soviet Union] in 1976, including 261 [sentenced] for anti-Soviet propaganda... 68,000 went through "prophylactic measures" [that year] that is those who had been summoned to the KGB and warned that their actions had been "unacceptable."[9]

In common practice, the state denied jobs, promotions, and opportunities to study in institutions of higher learning to many more thousands of people. With the KGB's watchful eye known to everybody, such routine actions of the authorities forced millions who were less courageous to conform. Communists always considered thoughtcrime unforgivable. These continuing atrocities did not diminish the enthusiasm of many fellow travelers and leftist intellectuals in the West who admired and supported the Soviet state and Marxist ideology.

The Marxist-Leninist lodestar required that workers and peasants (that is members of collective farms in absence of private properties), and not "rotten intellectuals," constituted a majority of the rank and file in the Communist Party. Correspondingly, they accounted for slightly more than one-half of the membership due to incentives and careful filtering and management of admission.

The Kremlin used these Party member workers and peasants as props and extras for the regime window dressing. The real power on all levels of the system rested in the hands of the ruling class. Consequently, career-oriented scientists, engineers, doctors, and other specialists lined up to join the Communist Party ranks. There were also many true believers, as the indoctrination began at an early age and continued through all stages of life. Contradicting the Party line was truly dangerous.

Similar to CPSU, workers, peasants, herders, and fishermen constitute today about 45% of the active members of the Chinese Communist Party. The real rulers of China propagate an ideological fiction that this composition of the CCP reflects the leading role of the working people.

It would be highly educational for assorted neo-Marxist intellectuals packing American university campuses and populating media to spend some time reading the foundational treatises of Proudhon, Engels, Lenin, Mussolini, Goebbels, Stalin, Mao Zedong (Tse-tung), and other socialist luminaries. This could help them better understand how these classics viewed intelligentsia in capitalist countries as well as foresee their inevitable personal destruction, often meaning physical extermination, when a social-justice paradise arrives.

9. Chernyaev, 2008, diary entry on January 3, 1976

In a letter to a leading Soviet writer, Maxim Gorky, Vladimir Lenin famously described intellectuals in capitalist countries as

> [bourgeoisie's] accomplices, the petty intellectuals, the lackeys of capital, who view themselves as the brains of the nation. In fact, they are not its brains but its sh*t.[10]

To join or not to join

In the Soviet Union, it was straightforward to become a Communist Party member for workers, military and police officers, apparatchiks, and the like. Or for somebody already in a position of authority in science, such as our laboratory heads Leonas and Baranov. When periodically approached and in order to avoid joining the ranks, the latter two responded, as an excuse, with invented stories of not being ready yet for the enormous responsibility of the membership in the Party, the vanguard of the working class, and similar nonsense.

For the rank-and-file scientists and engineers, especially in the Academy of Sciences, becoming a Communist Party member involved an effort because of quotas for "intellectuals." They competed against their ilk, against other like-minded scientists and engineers who wanted promotions and other benefits in life or be trusted with travel abroad. The latter constituted a major "carrot" in a closed society. Continuous pressure to conform and join the Party corrupted souls and served as an important tool in controlling the country.

Such an attitude to membership in a ruling communist party extended to all other socialist countries. A Czechoslovakian intelligence officer wrote, for example, that at that time "men join the [Communist] Party because they are ambitious, want to get ahead, and know that the membership of the Party is achieving that aim."[11] As a high-level Soviet functionary put it, "the party's support [for career advancement] can be compared to sugar in water: one couldn't see it but could constantly taste it."[12]

One day, Department No. 18 gained its first and only communist when my colleague Volodya Khromov joined the Party after patiently waiting in line and playing the required game. He did not have illusions about the Soviet system and was particularly knowledgeable about its inner workings. Khromov always wanted to become part of the administrative officialdom. The Party membership became an essential indispensable step for him.

---

10. Lenin, 1919
11. Frolik, 1976, p. 60
12. Katayev, 2015, p. 122

Privately, Volodya and I remained friends and open as before. We continued to make jokes and sarcastic comments about the rotten Soviet system, destructive ideology, fabricated history, leading personalities in the Kremlin, and Soviet stooges in the political left abroad.

The regulations required three Party members to form a local cell, a common rule for communists worldwide. The first Chinese taikonaut (astronaut) Yang Liwei even enthusiastically talked about a glorious day in the future when three Chinese communists form a cell at a space station orbiting the Earth.[13] As our department hopelessly lagged in Party membership, head of a communist cell of larger planetary science Department No. 4, Sergei Vasyukov, provided "ideological oversight" to us. Sergei was not an ambitious and vicious bone-headed ideologue as many others serving in such positions, stepping stones on their career paths. He typically did not bother us beyond what the Party Committee of IKI demanded from him.

One day, Vasyukov stopped by after working hours when our group celebrated the approaching New Year's Eve. Such events usually took place in the laboratory hall of Evlanov, next to a large vacuum chamber. Vasyukov eagerly took part in a few rounds of shots of alcohol mixed with water and accompanied by "let-us-have-a-small-drink" toasts and friendly banter. The institute issued alcohol for the maintenance of experimental facilities. Diverting such a "valuable state resource" to private consumption did not present a conflict to Party members, as well as to anybody else in the country (see Chapter 4).

Then Sergei Vasyukov pulled me aside for a private talk and volunteered to recommend to place me, "a nice and hard-working guy," in a long line for application to Party membership. The process would have taken several years to accomplish. I thanked Sergei for the "honor" but politely declined. Rejection of the proposition indicated at least a lack of enthusiasm about the political system and perhaps hidden hostility to the regime. Vasyukov became less friendly to me afterward but fortunately never turned into an enemy.

The absence of Communist Party members in the department did not make the environment or conversations much more open. After all, this was a country with abundant informers. The Party, media, and popular culture had encouraged, praised, and celebrated denunciations only a few decades ago, with many millions of human beings cheerfully exterminated.

IKI and Department No. 18 did not differ much from other physics establishments in the Academy of Sciences, a relatively well-informed

---

13. Xinhua, October 21, 2017

and "free-thinking" part of the society by Soviet standards. However, even those few who felt skeptical about the Soviet system and ideology had been trained from infancy by the totalitarian state to keep their mouth shut and behave to survive. They knew what to say, where, and to whom. In the close circles of their most trusted friends, they could boil with anger about injustices, stupidity, and the cruelty of socialism, but then with their tail between their legs, they acted "as required." Especially, if authorities cleared them to be in senior positions or for travel abroad.

Besides, the overwhelming majority of critically minded Soviet subjects did not have problems with the substance of socialism, a result of centuries of the country's oppressive history and decades-long Marxist-Leninist indoctrination and incessant propaganda brainwashing. The concepts of individual freedoms, representative democracy, and private property sounded abstract, while the ideas of unalienable rights, individual responsibility, a market economy, and limited government looked truly alien.

The people rather felt unhappy about the implementation of the Marxist ideology, obvious inefficiencies, unfairness, and cruelty of the system, and deviations from what they thought true socialism should be. A little more prosperity, openness, and transparency would have satisfied them, not unlike the ideas of Gorbachev when he tried to save the ideology later. Only very few viewed Marxism-Leninism, and more broadly socialism, as inherently destructive, dehumanizing, immoral, and perhaps evil.

One time in the mid-1980s I happened to be at a friendly party of Baranov's theoreticians. Occasionally, they invited me to their gatherings. Our conversation turned to the banned writings of Alexander Solzhenitsyn. I mentioned that one could easily bring his famous *Gulag Archipelago* from abroad. Actually, I had bought it in English translation at a flea market in Warsaw in brotherly communist Poland during my very first business trip to the Space Research Center in Warsaw.

"How dare you say this, Misha!" exclaimed a senior researcher with indignation. Then, she added as a friendly warning of a colleague experienced in life and accustomed to living in fear, "If they find it, they will never let you travel [abroad] again!" Privileges corrupt and tame.

Finding and confiscating the books by Solzhenitsyn and George Orwell were among the highest priorities of Soviet customs and border control officials for many years.[14] The state promptly and harshly punished the offenders.

---

14. e.g., Golovanov, v. 2, p. 325, 2001

Verboten travel and patriotic duty

Sometime in 1978, director of the *FOM Institute of Atomic and Molecular Physics*[15] (AMOLF) in Amsterdam, the Netherlands, Jaap (Jacob) Kistemaker and his deputy Joop Los visited our group. Los would soon replace retiring Kistemaker as director. At that time, AMOLF shined as a world-famous institution in atomic physics. It also had an extensive program of bringing visiting scientists from everywhere in the world, including the Soviet Union. The institute is no more. At the turn of the millennium, it completely changed its area of research and even its name.

Leonas had known Kistemaker and Los for many years and they were familiar with our experimental work in atomic collisions. By that time I had been on the IKI staff for a year and actively pursued the development of position-sensitive detectors based on microchannel plates, or MCPs (Chapter 5). It occurred to me during discussions with the guests that such detectors might exhibit a peculiar negative feature in measurements of differential scattering. After a quick back-of-the-envelope estimate, I brought the issue up. The effect certainly interested the visitors, but its magnitude was unclear. Nobody had ever thought about it.

In the evening, the Dutch colleagues left Moscow on a trip to the *Ioffe Physical-Technical Institute*[16] of the Academy of Sciences in Leningrad. (After the collapse of the Soviet Union the city reverted to its original pre-revolutionary name Saint Petersburg. I use its name from the Soviet times, Leningrad, throughout the book.) The institute, commonly known by the Russian abbreviation LFTI, standing for *Leningrad Physical-Technical Institute*, was a leading Soviet research organization engaged in fundamental research in atomic physics. The guests planned to return and visit IKI again in a few days.

With Kistemaker and Los away, I wrote and debugged a computer program simulating detector performance by the Monte Carlo technique. Yes, the effect showed up as predicted at small scattering angles.

Personal computers did not exist then, and one used mainframe machines such as IKI's BESM-6 and ES-1040,[17] the latter similar to IBM System/360. A group of technicians, all young women, took scripts of codes in a high-level language, such as Fortran, hand-written by scientists on specially formatted sheets of paper, and transferred them to punch cards

---

15. *FOM-Instituut voor Atoom- en Molecuulfysica* (AMOLF) in Dutch. FOM is a government foundation, *Fundamental Research on Matter*.
16. Named after physicist Abram F. Ioffe.
17. BESM stood for *Bol'shaya Elektronno-Schetnaya Mashina* (Large Electronic-Computing Machine); ES stood for *Edinaya Sistema* (Unified System).

2. Department of Space Gas Dynamics Mike Gruntman

(Fig. 2.6) on mechanical devices. A code thus turned into a stack of cards that were then fed into the computer.

The shift operators, also young women, assembled codes of many users into one package. They usually ran two such batches of programs each day. One batch started in the morning and was executed during the day. They launched the other batch in the evening for the night shift.

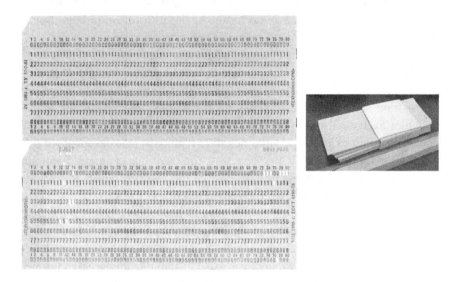

Fig. 2.6. Standard 7-3/8 in. by 3-1/4 in. punch cards introduced by IBM in the late 1920s. Each card represented one line of computer code, with 80 columns corresponding to 80 symbols. Each column of the card rendered a particular symbol in the binary code by a combination of holes (1) and their absence (0) in the set of 10 rows. Mainframe computers converted stacks of cards into programs and then compiled and executed them. Top-left: a blank card. Bottom-left: a card with the encoded word EJECT in columns 10-14 and the numbers 0001 9000 in columns 73-80. A stack of punch cards is shown on the right. Photographs courtesy of Mike Gruntman.

To debug my simple program, I needed to run it a couple of dozen times which would have taken two weeks. The mainframe computers were not particularly reliable, however. Something malfunctioned several times each day. The engineers then stopped the computer and fixed it. The operators restarted the remaining unprocessed programs after testing the machine by executing a small program from the batch. Making the operator choose one's code for such periodic testing allowed running and correcting it a few times during one day.

A romantic relationship with a computer operator would have done the trick easily. Since I was not friends with any, I bought several chocolate bars to get favorable treatment. Besides, I did my debugging during the weekend. Working in the institute on Saturdays and Sundays required special permission, and IKI was virtually deserted. The absence of competitors and the chocolate bars reinforced by flirting proved effective. I got the desired attention of the operator shifts.

When Joop Los visited us again the next week, I had my results ready. It was clear that the discrete microchannels of MCPs in position-sensitive detectors would affect the accuracy of intensity measurements in scattering on the smallest angles. My computer simulations proved the new effect conclusively.

Los was very much pleased with our acquaintance and interactions. Before departure, he unexpectedly told me that they would love to bring me as a visiting scientist to AMOLF. (The Dutch institute covered all expenses except for airfare.) Los realized, however, that the officials would probably never permit me to travel.

In a couple of months, the AMOLF director sent me a formal invitation to spend three months as a visiting scientist in the institute in Amsterdam. This invitation quickly and firmly defined me as someone not allowed to travel abroad. My close colleagues and direct peers Kalinin and Khromov spent their "standard" three months each in AMOLF as visitors. My travel was verboten.

The travel restriction was not unexpected. Not only I did not belong to the Communist Party, but I didn't even pretend to be interested in joining it. No doubt my name had already appeared in reports of informers. Even my boss felt uneasy about the possibility of me going to Amsterdam, as his job and livelihood, understandably, would be on the line as well. "You have to promise me that you would not do anything stupid," Leonas told me, knowing that I could defect. Making such a promise to a close senior colleague responsible for me would have forced me—*noblesse oblige*—to keep it. The totalitarian system effectively made everybody hostage.

Being single meant leaving no hostages back home to assure a return to the beloved socialist homeland. And perhaps worst of all, the state put Jews in a "special" category. On top of that, my Germanic last name belonged to another singled-out group, as icing on the cake. Consequently, I experienced the wonders of real socialism, which treated people as members of groups

2. Department of Space Gas Dynamics   Mike Gruntman

rather than individuals. Ironically but not surprisingly, neo-Marxists push the United States in the same direction today, away from merit, by advancing identity politics and diversity by exclusion.

If I were permitted to visit AMOLF in 1979, I would have become the youngest visiting scientist in the history of this famous institute. I did not know whether this statement was precise or perhaps just "close enough" to the fact, but this was what institute directors Kistemaker and Los told me in Moscow. The trip did not happen anyway.

Later, my Polish colleagues exerted pressure on IKI and the Academy of Sciences to let me visit Warsaw. By that time, in the early 1980s, I had been leading collaborative work on energetic neutral atoms which made my travel truly indispensable for the effort. "Fraternal" Poland was not the Netherlands and one could not escape to the free world from there. So, the Academy finally authorized my travel to this tightly guarded barrack in an unhappy Marxist concentration camp. Then the country was under martial law and the Polish government drove the outlawed independent Solidarity trade union to the underground. Consequently, those trips led to some "extracurricular" activities on my part (Chapter 9).

I finally visited AMOLF in 1988. The first thing that I heard from the institute director after his warm welcome greeting was that they would not help me in any way or form if I decided not to go back. They could not and would not risk their well-established working relations with the Soviet science establishment. Pure business considerations outweighed personal sympathy and political views. Not only were my family and my boss hostages in Moscow, but the totalitarian state effectively forced the Western partners to play along. I sincerely thanked Joop Los for being frank—I truly appreciated the openness—and assured him, and I meant it, that I would not place any burden on the institute in case of such developments.

Cooperation of businesses, nonprofit organizations, and other entities of the free world with totalitarian states was and is most common. Like the Soviet Union in the past, today, the People's Republic of China projects its economic power and advances the interests of its Communist Party in many countries through an army of willing or pragmatic advocates.

Scientists have always been a primary target in such influence operations, an easy target for manipulation when a socialist ideology is involved. Long, long ago in 1933, prominent European scientists did not want well-known Soviet physicist George Gamow to stay in the West, "[Y]ou cannot do this. You have to go back to [Soviet] Russia," he was told.[18] Gamow's

---

18. Gamow, 1975, pp. 128-130

particular situation and circumstances were different. In my case, AMOLF's practical, programmatic considerations—not ideology—drove its attitude.

For the first time, I crossed the Iron Curtain in December 1987, a few months before the visit to AMOLF. Despite some relaxation of travel restrictions introduced by *perestroika* (restructuring) and *glasnost* (openness) by that time, the Academy of Sciences seemed to let me travel for a few days by mistake, due to a glitch in the cumbersome and complacent control system. On this my first trip to West Germany, one senior scientist, full of adoration for Gorbachev, passionately urged me to go back to Moscow in words similar to what Gamow had heard decades earlier. At the same time, another colleague, Hans Fahr (Chapter 8), privately assured me of his help in connecting with the proper authorities if I decided to "disappear into the fog" in West Germany. (Thank you, Hans.)

Now, back to IKI and Moscow in 1979. Based on the analysis of the position-sensitive detector performance and discussions with Joop Los, I wrote my first article[19] in English and published it in *Journal of Physics E*, a leading foreign journal[20] in scientific instrumentation and experimental methods in physics. Publications abroad required obtaining numerous approvals from IKI and the Academy of Sciences, including from security officials.

Two administrative units with overlapping responsibilities oversaw matters related to security in Soviet research institutions. The *Secret Department* of IKI focused on and managed classified documents, while the *Regime Department* concentrated on people. The local Communist Party committee and institute administration assisted the Regime Department in its mission of people control. No doubt the department and other external bodies associated with secret police also maintained informers among IKI employees, including scientists. The East German Ministry for State Security, the *Stasi*, so well known in the West for its extensive network of informers, had not invented the wheel and learned much of the craft from their older Soviet brothers. The totalitarian country practiced this art of snitching for generations, and scientists, students, and even the military officers were not exempt in any way.[21]

One day, head of the Regime Department of IKI summoned me for a "conversation." The manuscript of my article was on his desk. Stern, unfriendly, and menacing, he gave me a long indoctrination session about "the patriotic duty of Soviet scientists to publish in Soviet scientific journals."

---

19. Gruntman, 1980a.
20. published by the Institute of Physics in the U.K.
21. e.g., Orlov, 1991, pp. 85-88, 105, 106, 131, 132

After that, he signed off on the permission to send the article to a journal in Great Britain. Ironically, neither he nor I knew at that time that I had already "published" in the West without any approval, as explained in the next chapter.

Working on scientific problems was certainly much more pleasant than dealing with the security apparatus.

3. Blazing My Own Trail

Research fellow

As my boss Vladas Leonas had promised, the institute changed my engineering staff position to a junior research fellow after one year. In fact, I worked as a scientist from day one. Therefore, it was purely a question of a formal title adjustment as well as a 10-ruble increase (9%) in my monthly salary.

The average salary in the national economy was about 140 rubles per month in the mid-1970s, with entry-level engineers paid 110-120 rubles.[1] The official exchange rate of the ruble was close to one and a half U.S. dollars, while on the black market one dollar bought 5-6 rubles at that time.[2]

Every newly hired graduate, a "young specialist," underwent evaluation after one year of work. My record was excellent, reflecting diverse activities and boiling energy. I did not hesitate to tell the evaluation committee about the desire to change my staff position from an engineer to a research fellow. One senior member of the committee, Vladimir G. Kurt, suddenly began to explain to me the virtues and honor of being an engineer. I could not believe it. It took an effort to listen to this sermon that lasted three minutes. It was not easy. With restraint, I politely repeated my request to be reclassified as a research fellow. The committee did not have any objections at all and unanimously approved the recommendation. The administration soon changed my title.

There were major differences between Kurt and me in our positions in the institute and age. At the same time, our scientific interests overlapped

---
1. Srednie zarplaty..., 2021; Gruntman, 2011
2. also, Gruntman, 2011

a little bit in heliospheric neutral atoms. With time, we established excellent professional relations. Kurt would later serve as an official reviewer of my Ph.D. thesis.

At the beginning of the fourth student year in 1974, Leonas suggested exploring possible ways of measuring electric fields in space from a spacecraft using electron and ion beams. The true interests of my boss were in atomic collisions. His laboratory was at IKI, however. Naturally, he looked for ways to apply its atomic-beam expertise and associated techniques to space experiments.

It did not take me long to come up with an original but rather complex concept for measuring electric fields. Because of inexperience, my imagination ran wild and I "invented" many untried approaches. It was certainly fun since nobody provided any guidance and expertise and nothing restrained my pursuit.

Interestingly, many years later, I read in the literature about some elements of the techniques and approaches I had come up with. Their independent applications in unrelated areas validated the old ideas. My concept of measuring electric fields did not look practical, however. It required a multi-year effort to explore and prove, or disprove, its feasibility in the laboratory. Consequently, we decided to abandon the project.

Detection of interstellar helium atoms in ... 1975

Early in my fifth student year in 1975, Leonas and I decided to look into the possibility of direct detection of individual neutral atoms in space. Our group had registered the atoms in laboratory experiments, although the energies and species were different.

So, why not in space?

The Sun moves with a velocity of about 25 km/s, or 5 AU/yr,[3] with respect to the surrounding interstellar medium, known as the *local interstellar medium*, or LISM. The motion of interstellar gas and plasma relative to the Sun is often described as interstellar wind. As the number densities of the partially ionized interstellar gas and the solar wind plasma filling interplanetary space are very small, interstellar helium atoms fly into the solar system practically without collisions.

The Sun's gravity further accelerates the incoming interstellar atoms. Physical estimates showed that the helium atoms should have kinetic energies exceeding 100 eV relative to an observer on a spacecraft moving

---
3. One astronomical unit, 1 AU = 1.496×10^{11} m (149.6 million kilometers), is the mean distance between the Sun and Earth.

with the Earth around the Sun. Particles with such energies could knock out electrons from surfaces, the phenomenon known as electron emission. One can then detect the emitted electrons using devices known as *secondary electron multipliers*, or SEM. Therefore, these knocked-out electrons could open a way for detecting and counting individual interstellar helium atoms hitting the exposed sensitive surface of a sensor. Possible detection efficiencies of such 100-eV atoms were unknown. Even more importantly, suppression of the expected huge noise count rates of secondary electron multipliers due to background heliospheric ultraviolet (UV) and extreme ultraviolet (EUV) photons looked unrealistic. This was the hopeless situation in 1975.

At that time, nobody in the world had directly registered individual neutral atoms in interplanetary space. I had never even heard then about anybody interested in the direct detection of such atoms in situ. In fact, not that many had an interest in neutrals in general.

In our department, Vladimir Baranov theoretically explored a global interaction of the interstellar wind with the solar system where neutral gas played a role. He applied gasdynamical concepts to describe plasma flows. The marginal overlap with Baranov's interests would strengthen with time and evolve into a scientific collaboration and friendship.

Several groups in the world had been measuring solar photons scattered by atoms in interplanetary space since the 1960s. In IKI Vladimir Kurt headed one such active and accomplished laboratory with a solid science record. As a true astronomer, he inhabited a world of ultraviolet photons and, as it seemed to me, looked with some sincere condescension on my attempts to directly detect individual neutral atoms in the heliosphere. My unconventional approach appeared hopeless at that time, which only reinforced the primacy, or orthodoxy in my view, of observing neutral atoms through detection of scattered solar photons.

These photons excited the resonant transitions of interplanetary atomic hydrogen (wavelength 1216 A or 121.6 nm) and helium (584 A or 58.4 nm) and constituted the biggest source of noise preventing direct detection of neutral atoms by secondary electron multipliers. Consequently, I continuously looked for ways to suppress the effects of the photons on sensors and achieved some successes. The conceived and advanced approaches would become indispensable for ENA instruments in the future.

The early exploration of ways to detect interstellar helium at IKI in 1975 became my first step on a road leading to ENA imaging.

Secondary electron multipliers

When an incident particle or a photon knocks out an electron from a surface, a secondary electron multiplier can pick it up and convert it into an avalanche of millions of electrons. Electronics can then reliably register such bursts of electrons and thus count individual incident particles.

In the mid-1930s, Vladimir Zworykin of the *Radio Corporation of America* and John R. Pierce of *Bell Telephone Laboratories* developed the first discrete dynode electron multipliers in the United States for use in various applications.[4] Leonid Kubetsky had built conceptually similar devices for his laboratory experiments even earlier in the Soviet Union.[5]

An electron entering a secondary electron multiplier is accelerated by a static electric field and hits the first dynode, producing a few secondary electrons. These electrons, in turn, are further accelerated and hit the next dynode, producing more secondary electrons. The number of electrons thus increases exponentially. After a dozen or more multiplication stages, one obtains a burst of many thousands or millions of electrons at the output of the device. Such multidynode electron multipliers could be rather bulky, for example, one inch in diameter and four inches long as illustrated by VEU-1 manufactured in the Soviet Union (Fig. 3.1). Here VEU stands for *vtorichnyi elektronnyi umnozhitel'* in Russian, or secondary electron multiplier.

Fig. 3.1. Typical Venetian-blind discrete dynode secondary electron multiplier VEU-1 for operation in a vacuum. Particles enter the multiplier at the left end; the dynodes are in the middle. Photograph courtesy of Yu. V. Gott.

Nuclear physics and other fields have widely used SEMs for decades. In a major advancement of the technique in the late 1950s and early 1960s, George W. Goodrich and William C. Wiley of the Research Laboratories

---

4. Zworykin et al., 1936; Pierce, 1938
5. Kubetsky, 1937

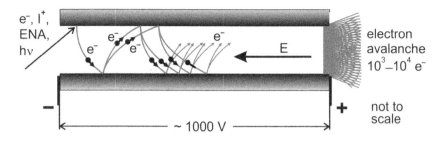

Fig. 3.2. Schematic of a channel electron multiplier, CEM, detecting and counting individual energetic particles and photons in a vacuum.

Division of the *Bendix Corporation* showed that efficient electron multiplication could be achieved in channels.[6] This breakthrough finding led to a *channel electron multiplier*, or CEM, that is basically a tube made of highly resistive but slightly conducting glass.

A high voltage is applied across the ends of such a tube where it creates an electric field E in the axial direction (Fig. 3.2). An incoming energetic electron, ion, neutral atom, or ultraviolet or X-ray photon knocks out the first electron at the channel entrance. The axial electric field then accelerates the electron to an energy of 50-150 eV and it collides with a channel wall, producing secondary electrons. The field then accelerates these electrons until they collide with the wall. The process repeats itself many times, resulting in an avalanche of many thousands or millions of electrons exiting the CEM.

CEMs are electrostatic devices where the total applied electric field, rather than its gradient, determines the acceleration of electrons. Consequently, the multiplication properties of a channel electron multiplier depend on the length-to-diameter ratio of the channel rather than on its absolute size. This feature opens a way for miniaturization, particularly to combining millions of small straight channel multipliers into the so-called microchannel plates (Fig. 3.3). An early CEM patent by Goodrich and Wiley envisioned "a plurality of [channel] multipliers ... arranged to intensify a light image."[7]

Originally developed for image intensifiers in night-vision devices,[8] MCPs not only enabled reliable detection and counting of individual particles and photons[9] but also provided the basis for position-sensitive detec-

---

6. Goodrich and Wiley, 1962
7. Goodrich and Wiley, 1960
8. Morrow et al., 1988; Schnitzer, 2000; Bender, 2000
9. e.g., Colson et al., 1973; Leskovar, 1977; Wiza, 1979; Lampton, 1981; Gruntman, 1984; Timothy, 1985

3. Blazing My Own Trail
Mike Gruntman

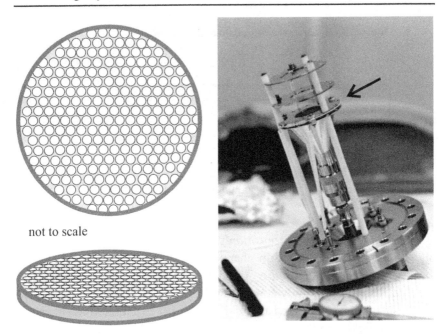

Fig. 3.3. Schematic of a microchannel plate with millions of small identical straight channels in a hexagonal pattern, each acting as independent channel electron multipliers (left). An MCP converts incoming particles into localized bursts of electrons at its exit where electronics amplify and count the bursts as detected individual events. Typically, two or three MCPs are mounted in series to provide the desired electron multiplication.

The photograph on the right shows (arrow) such a detector. MCP sensitive areas vary from one inch to a few inches in diameter; diameters of microchannels could be from 6-20 μ. (1 μ = 10^{-3} mm = 10^{-6} m is one micron.) An MCP detector converts an incoming particle into an electron avalanche locally, at a place where it hits the sensitive surface of the front MCP. This feature opens a way for building detectors that not only register individual particles but also determine positions of the electron avalanches at the MCP exit and thus the positions (co-ordinates) of the corresponding impact points of the registered particles on the sensitive surface. Such sensors are known as position-sensitive detectors, PSDs.

MCPs also form the basis for passive night-vision devices relying on starlight illumination. A photocathode near the MCP entrance converts an incoming photon into a photoelectron that, after acceleration, enters the MCP and produces an electron avalanche at the plate exit. Then, the electrons of the avalanche are further accelerated and strike a phosphor screen, with a resulting flash of light. Such night-vision systems operate in a vacuum in sealed enclosures and amplify image brightness by a factor of ten thousand and more.

Photograph courtesy of Mike Gruntman.

tors. Such latter devices determine a position, the coordinates, of the impact of each individual registered particle at the detector's sensitive surface.

Various types of secondary electron multipliers, including CEMs and MCPs, would be widely used in physics laboratories across the world. By 1973, the Soviet industry had begun producing compact open-type (that is operating in a vacuum) channel electron multipliers. The laboratory of Leonas widely used such CEMs with a 10-mm diameter funnel and spiral channel, VEU-6 (Fig. 3.4), for particle detection in collision experiments.

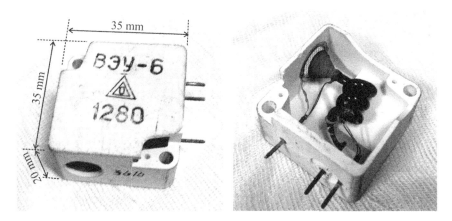

Fig. 3.4. Commercially available channel electron multiplier with funnel VEU-6 capable of detecting and counting individual energetic particles and photons in a vacuum. The funnel increases the sensitive area of the detector. The curved spiral channel efficiently suppresses ion feedback. Photographs courtesy of Yu. V. Gott.

From 1975, my work focused on channel electron multipliers and microchannel plates important for experiments in the laboratory of Leonas. At the same time, I hoped to apply these detectors for the registration of individual neutral atoms in space. Other members of the group shared an interest in detectors for studying collisions, but not neutral atoms in space. Here I was completely on my own.

Secondary electron multipliers reliably registered individual neutral atoms with energies of 600 eV and higher in our laboratory experiments. No physical reasons fundamentally precluded the detection of interstellar helium atoms with the expected 100-eV energies. The detection efficiency of SEMs, however, rapidly drops as energy decreases. So, it would be much smaller for interstellar helium atoms but nobody knew how small.

Master's thesis

My "own" science interests emerged in the early student days. They deviated from the atomic collisions of my advisor and his laboratory and focused on the physics of neutral atoms in space plasmas; secondary electron multipliers; and detection of various individual photons and particles, especially neutral atoms in laboratory and space. The experimental part included such diverse areas as vacuum technology; particle collisions with surfaces; secondary electron and ion emissions; interaction of particles with matter (scattering, energy loss, straggling, charge state); sources of electrons, ions, and photons; monoenergetic neutral atom beams; and nuclear-physics-type electronics.

My MFTI master's thesis focused on interstellar helium fluxes at the Earth orbit. The thesis theoretically considered expected properties of the fluxes at 1 AU from the Sun as seen by an observer moving with the Earth (that is on a spacecraft near the Earth) and predicted their directional

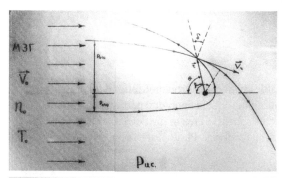

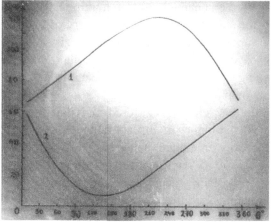

Fig. 3.5. Slides from the presentation at the defense of my master of science thesis in May 1977.

Top: direct and indirect trajectories of interstellar helium atoms entering the solar system.

Bottom: energy of helium atoms (in eV) following these two trajectories as seen by an observer moving with the Earth around the Sun as a function of the position (angle θ in degrees) along the Earth orbit. The angle θ is measured from the upwind (interstellar wind) direction.

The Earth rotates clockwise around the Sun in the figure.

intensities, conveniently characterized by "perpendicular temperatures." Measuring such characteristics in a space experiment would have determined the velocity vector, temperature, and number density of neutral gas in the interstellar medium surrounding our Sun.

Those familiar with fluxes of interstellar neutral atoms entering the solar system would certainly recognize the basic concepts and dependencies in crude figures (Fig. 3.5) from the thesis defense in 1977. One can see a sketch of helium atom "direct" and "indirect" trajectories in the gravitational field of the Sun and the corresponding atom energy dependences for these two trajectories on the position (angle theta in degrees) of an observer moving with the Earth around the Sun.

The experimental part of the thesis focused on the performance characteristics of VEU-6 registering individual ions and atoms with energies of several hundred electronvolts and higher. Our experimental setup fundamentally limited the range of possible particle energies. Therefore, it was impossible to try the detection of atoms with energies below 600 eV. I defended the thesis in May 1977 and graduated with a physics degree.

In 1990, fifteen years after beginning my lonely work on interstellar helium flux detection, a West German instrument on the *Ulysses* spacecraft, directly registered such fluxes for the first time. The sensor relied on secondary ion emission from its sensitive surface.

What I did not know during my student years in the mid-1970s was that a couple of visionary scientists, as described below in Chapter 8, had begun a pursuit of direct detection of interstellar helium only a few years earlier in 1972. Their effort would ultimately lead to this successful pioneering experiment on Ulysses twenty years later. Another space instrument on NASA's Interstellar Boundary Explorer would measure interstellar helium atoms fluxes again in 2009.

My first publication was ... a NASA TM

The experimental work for the master's thesis led to my very first published journal article[10] (with a coauthor, Kalinin). It appeared in the leading Soviet physics instrumentation journal, *Pribory i Tekhnika Eksperimenta*, or PTE. In those days, the Plenum publishing house in the United States had been translating this journal cover to cover into English under the title *Instruments and Experimental Techniques*. Today, the journal (ISSN 0032-8162) is active in Russia, and Springer publishes its English translation.

Our article became one of the very first publications, perhaps even the

---
10. Gruntman and Kalinin, 1977b

first, in the Soviet physics literature, on the VEU-6 characteristics detecting energetic ions and neutral particles. Earlier, the developers of CEM technology in the industry had described its properties for detecting electrons.[11]

Writing an article provided an important "educational experience." A journal copy editor poured red ink on the manuscript, cleaning up the style and language. This embarrassing but necessary drubbing caused me to step up efforts in polishing the manuscripts. The communication abilities of Soviet science and engineering students and recent graduates did not differ much from those of their present-day American peers. That is to say that they were and are truly lousy. The only remedy here was and remains to follow old advice, paraphrasing, "Flogging will continue until writing improves." And, I would also add, in the style of Admiral David Farragut, "Damn the feelings!" Snowflakes would never excel in communicating in the merit-based competitive environment of physics.

The first journal article gave a reason for celebration, but my formal first publication was a preceding IKI preprint. Actually, it turned out also to be, unbeknownst to me at the time, a technical memorandum by the United States government!

IKI had a well-established system of publishing its reports, called "preprints," which was common for many leading physics research institutions across the world. The institute typically printed one hundred copies of such preprints and sent them to main national libraries and libraries of research organizations. The preprints were entered into catalogs, databases, and publication records of the authors. They underwent IKI's internal quality and patentability reviews and the officials cleared their content from the point of view of secrets and ideological purity.

In common practice, many authors first rapidly published a preprint and then sent a compressed version to a physics journal. Leading science journals required concise publications and discouraged verbose manuscripts. It took more than one year for an article to appear in a journal, while a preprint could be issued in several months. The speed was sometimes important for priority. A preprint could also include many technical details omitted in articles.

Before sending our manuscript to PTE, Kalinin and I published its expanded version as Preprint 311[12] of IKI, detailing the study of the performance characteristics of VEU-6. This was my very first official publication on a narrow technical issue, obscure and of interest and known to very few. Or, so I thought.

Many years later, in 2019, I discovered that the U.S. government had

---

11. Ainbund et al., 1974a,b
12. Gruntman and Kalinin, 1977a

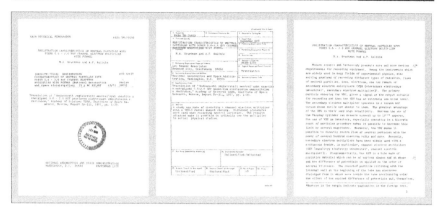

Fig. 3.6. The first three pages of NASA TM-75540, a translation of IKI Preprint 311, published in 1978. For more than 40 years, I did not know about the existence of my very first publication in the United States.

GRUNTMAN, M. A. 000000303628
//ISSLEDOVANIYE KHARAKTERISTIK DETEKTORA-ENERGOANALIZATORA BYSTRYKH ATOMOV H NA OSNOVE FOL'GI, ACADEMY OF SCIENCES USSR, INSTITUE O
F SPACE RESEARCH//
00/00/1981,V----,N----, PP 0001-0065
REPORT PR-667
STUDY OF THE CHARACTERISTICS OF A DETECTOR-ENERGY
ANALYZER FOR FAST H ATOMS, BASED ON A FILM
*NASA TM-76835

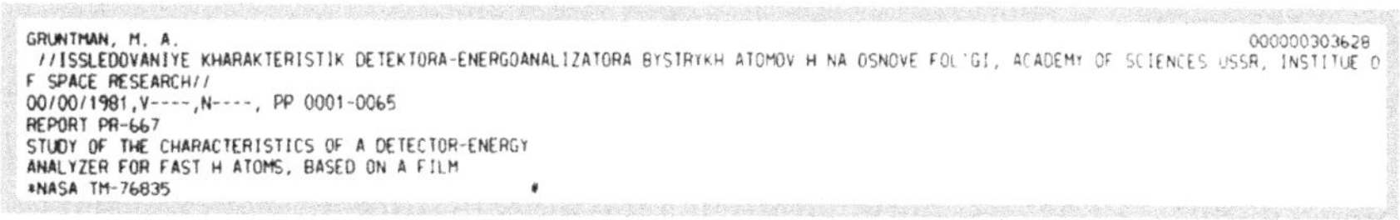

Fig. 3.7. Preprint 667 (listed as Report Pr-667), IKI, 1981, translated and published as NASA TM-76835, in a CIA database.

noticed this preprint, translated it into English, and published it as a NASA technical memorandum (TM), NASA TM-75540[13] in 1978.

My very first publication became a NASA TM!

Figure 3.6 shows the first three pages of this TM-75540. A government contractor in California did a lousy job and delivered an awful translation.

Further searching also revealed that this was not my only IKI preprint translated by the U.S. government. The recently declassified Central Intelligence Agency's (CIA) database of translated Soviet scientific and technical materials lists at least one other preprint, Preprint 667,[14] published as NASA technical memorandum TM-76835 in 1981 (Fig. 3.7). Amusingly, NASA denied my formal request under the Freedom of Information Act (FOIA) for a copy of TM-76835 because ... its copyright belonged to IKI. A NASA official even suggested contacting IKI directly.

So, when the head of the IKI regime department lectured me on my

13. Gruntman and Kalinin, 1978

14. Gruntman and Morozov, 1981.

patriotic duty to publish in Soviet journals, my very first publication quietly appeared as a NASA technical memorandum in the United States without his august approval.

Who knew in those dark hopeless days of the seemingly unshakable totalitarian state, akin to another eternal *Tausendjähriges Reich* (Thousand-Year Reich), that this first "NASA publication" would become a small step on a long journey to the free world, leading to participation in NASA missions and founding a space engineering department at a university in the heart of the U.S. space industry. As a colleague in a senior position at NASA kindly commented on my TM discovery, "You were and are part of the NASA team. ... It is great you found this [translated publication] and should be really proud."

In the mid-1970s and a few years behind the United States, the Soviet industry began producing microchannel plates, the wafers with millions of tiny straight-channel electron multipliers (Fig. 3.3). Similar to the U.S., applications in the second and then third generations of night-vision devices drove MCP development, which provided a "major breakthrough" in imaging technology.[15] (More than 150,000 MCP-based image intensifier systems had been procured and deployed by 1987 in the armed forces of the United States alone.[16]) The first publication in a physics instrumentation journal on Soviet MCPs by their developers appeared in 1975.[17] We got our hands on MCPs in 1976 or 1977. At first, the manufacturer provided non-operational, rejected-in-production MCPs, which allowed me to start designing detectors based on these new promising secondary electron multipliers.

In a few years, we would be among the very first physics groups in the country to design, build, and successfully demonstrate MCP-based detectors for the registration of individual particles and photons.[18] Our understanding of performance characteristics of MCPs and CEMs was often ahead of such powerhouses with advanced physics culture as research institutes in nuclear physics and high-energy particle accelerators.

My master of science thesis sketched a naive possible approach to measure directional dependences of interstellar helium intensities using an MCP detector with position sensitivity, allowing the determination of coordinates of each registered individual particle. Such position-sensitive detectors, or PSDs, fundamentally differed from the so-called frame detectors based on charge-coupled devices, or CCDs. The latter first matured in

---

15. Ponomarenko and Filachev, 2007, pp. 135, 136
16. Morrow et al., 1988
17. Bragin et al., 1975
18. Gruntman and Kalinin, 1980

the United States for electro-optical imaging systems of reconnaissance satellites in the 1970s. They are ubiquitous today in the digital cameras of mobile telephones and laptop computers.

MCP-based PSDs were an emerging uncertain technology in the 1970s, with none in existence in the Soviet Union. Such position-sensitive detectors promised to dramatically improve the efficiency of differential scattering measurements in laboratory experiments. Consequently, my advisor Leonas became interested in them. PSDs could enable concurrent measurements of particles scattered in the entire range of angles of interest and thus replace a mechanically scanning "point" detector (a CEM with a pinhole aperture). They could also make practical the detection of particles at large scattering angles where intensities significantly dropped. In space research, photon- and particle-counting imaging detectors promised to lead to new generations of highly capable instruments in space plasma physics and extreme ultraviolet and X-ray astrophysics.

The transition from single microchannel plates employed in image intensifiers to detectors counting individual particles and photons was not simple. It required the use of sets of plates installed in series, as first demonstrated in Bendix Research Laboratories in 1973.[19] The next step of achieving position sensitivity would take much more effort. By the mid-1970s, very few physics groups in the world had succeeded in demonstrating basic, simple PSDs. And we at IKI did not have a single operational MCP yet.

This was the state of affairs when I jumped on the opportunity to develop position-sensitive detectors, my long-term strategic goal since 1976. One had to start first with basic MCP detectors for counting individual particles. With time, we would become the first and only physics group in the country, based on published physics literature, to build PSDs (Chapter 5).

Detecting individual neutral particles

In the 1970s, the efficiency of producing electrons in an impact of an interstellar helium atom on a sensitive surface remained unknown and unpredictable. Our atomic physics experience informed us how to design and build a required experimental facility with collimated monoenergetic beams of 100-eV atoms. Unfortunately, there were no resources to do this. The literature also did not provide any hard data on the registration of low-energy atoms.

The first reliable article on possible detection efficiencies of such atoms appeared only in 1982. Sam Cohen and a colleague at the *Princeton Plasma*

---

19. Colson et al., 1973

Physics Laboratory measured electron emission from a surface bombarded by neutral deuterium atoms with energies down to 50 eV.[20]

This important publication experimentally confirmed a practical possibility of using secondary electrons for the detection of 100-eV atoms. The efficiencies would obviously be small but not zero. The challenge of suppressing a superior ultraviolet background, however, remained. It was reassuring to know that somebody else in the world was interested in detecting low-energy atoms. By that time, I had already met Helmut Rosenbauer, who worked on this problem in West Germany (Chapter 8). The lonely days in neutral atoms were coming to an end.

The physics world is large but also small. In 2002, Sam Cohen and I happened to serve on a panel reviewing one government program. Sam was surprised and pleased to learn that his 20-year-old article had played such an important role in providing essential psychological support for my work.

As I could not explore experimentally detection of 100-eV atoms in my laboratory, the interest shifted to other more energetic populations of neutral atoms in space, such as a neutral component in the solar wind. Already in 1979, I firmed up the estimates of the fluxes of the neutral solar wind, or NSW, at 1 AU from the Sun[21] and made the development of instrumentation for their detection a priority.

Such ENAs with energies of about 1000 eV would efficiently produce secondary electrons at surface impact. Importantly, they can penetrate through ultrathin foils, which is accompanied by electron emission as well. Detection of NSW atoms also presented another formidable challenge: pointing the sensor only a couple of degrees away from the Sun, an exceptionally bright source of interfering energetic photons. This called for an efficient baffle.

The shift of interest to the detection of higher energy atoms prominently included registration of heliospheric ENAs originating at the interstellar boundary of the solar system, 100 AU away from the Sun. It began a 30-year journey, from the late 1970s to 2009, on a road to detect these atoms. Launched in 2008, NASA's Interstellar Boundary Explorer finally mapped the interstellar boundary in ENA fluxes for the first time (Fig. 3.8).

In 1983, the summary of our prior theoretical work and instrument development outlined a comprehensive space experiment[22] to detect neutral atom fluxes: neutral solar wind, ENAs from the interstellar boundary of the solar system, and interstellar helium fluxes. One section (4.4.4 *"Reflected"*

---

20. Voss and Cohen, 1982
21. Gruntman, 1980b
22. Gruntman and Leonas, 1983

Solar Wind) specifically focused on heliospheric ENAs from the solar system frontier 100 AU away. By the time of the publication, we had been working on the concept for several years. This comprehensive neutral atom experiment would become known as the GAS experiment[23] (Chapter 9). Note that the West German experiment on Ulysses to measure interstellar helium fluxes was also named *GAS*, causing some confusion.

Interestingly, nobody has definitely measured and experimentally characterized the properties of the neutral solar wind atoms until this day (see also Chapter 9). At the same time, nobody doubts their existence and the important role these ENAs play in the global interaction of the solar system with LISM.

Fig. 3.8. Cover of *Science* magazine with the IBEX mission mapping the interstellar boundary in ENA fluxes, 2009. Cover credit: American Association for the Advancement of Science.

Also by the early 1980s, I had discovered that the neutral solar wind would enter the surrounding interstellar medium and significantly disturb it (Chapter 7) up to distances of several hundred astronomical units.[24] An IBEX team later added new features to this effect, which provided a foundation for a tentative explanation of the so-called "ribbon" observed in ENA maps. Besides, the neutral solar wind is an important driver for the requirements to the science payload of NASA's planned *Interstellar Probe* mission to experimentally explore in situ the interstellar medium surrounding the solar system.

Behaving oneself vs. having fun

My colleagues in the high-energy collision group, Sasha Kalinin and Volodya Khromov, had been measuring the differential scattering of atomic beams for years. This was a traditional, "classical" experimental physics

---
23. Gruntman et al., 1990c
24. Gruntman, 1982

area that involved clean, high vacuum, collimated monoenergetic neutral particle beams and their scattering on gas targets. A "point" detector then scanned across the scattered beam, registering neutral atoms. A microcomputer partially controlled the experiments, which was a major achievement in those days.

A straightforward road to writing and defending a Ph.D. thesis included measuring the differential scattering cross-sections of a few beam-target neutral atom and/or molecule pairs, reducing experimental data to science results (primarily obtaining interatomic interaction potentials), and perhaps some improvements in experimental techniques and set up.

Five years older than I, Volodya Khromov had just received his Ph.D. degree. He advised me in a friendly manner, "If you behave yourself and follow the boss, then you can defend the thesis in three years [after graduating with the master's degree]. If you argue, snarl back, and push your own ideas, then it will drag for 7-8 years."

Volodya turned out to be right. I spent one year measuring the scattering of atomic oxygen on atoms of neon and argon. It was not challenging enough. Somehow, incremental advances never interested me in the past and do not ignite enthusiasm today. I always wanted to work on tasks that were never attempted and with unpredictable outcomes, where nobody knew how to proceed.

Consequently, my interests had shifted by that time to a much more exciting but seemingly hopeless direct detection of individual neutral atoms in space. Nobody knew much or cared about them. I steadily progressed with the development of MCP-based detectors for ENA imaging. During those early years, the development of position-sensitive detectors, another challenging area without established paths, dominated the work. Possible applications of PSDs in various fields led to many interesting professional contacts and "distractions," which made science life diverse and exciting. As a result, I received my Ph.D. degree 6.5 years after graduating from MFTI. Such a time frame was actually common for many young staff scientists in the Academy of Sciences and nuclear physics research centers in the country.

Defenses of Ph.D. theses took place in special science councils. Each council comprised one or two dozens of leading scientists. The process starkly differed from the American practice of defense in an essentially "family environment" at a committee composed of the Ph.D. thesis adviser and a few university colleagues.

The science councils appointed two official reviewers for each Ph.D. thesis, called "opponents," who provided an independent assessment of the presented theses. The word "opponent" did not imply an adversarial attitude but rather an independent qualified evaluation. In addition, the council requested a formal review of the defended thesis by another research institute specializing in the field. The latter review involved a few leading experts of that organization, and the candidate also gave a talk at a science seminar there. All these external evaluations and reviews that were also followed by the final approval by a national board provided the quality control of Ph.D. degrees.

IKI had a few science councils authorized to conduct Ph.D. defenses. I defended my thesis at the science council headed by IKI director Roald Sagdeev. More than a dozen of the most senior IKI scientists served on his council, which usually dealt with defenses of the higher level D.Sc. degrees. This latter degree roughly corresponded to the habilitation qualification (a post-doctoral degree) in Germany, France, and some other European countries. In exceptional cases, Sagdeev's council also conducted defenses of "regular" Ph.D. theses, and mine fell into that category.

My thesis "Substantiation and Development of Methods for Detecting Neutral Particle Fluxes in Interplanetary Space" (Fig. 3.9) covered the physics areas more diverse than a typical Ph.D. and included subjects that were rarely mixed. Suffice it to mention that the two appointed official reviewers, the opponents, were Vladimir Kurt and Yuri Gott, representing the fields far apart.

The reviewers had to be experts in the areas of the defended work. Astronomer Kurt headed a laboratory at IKI and specialized in interstellar gas and ultraviolet optical space instruments. An experimental physicist from the *Kurchatov Institute of Atomic Energy*[25] (IAE), Gott, excelled in studies of particle interactions with matter and their applications for fusion plasma research, a classical area in nuclear physics. He also authored a unique monograph[26] in this field in 1978. IKI assigned the Physical-Technical Institute in Leningrad to serve as the official reviewing organization. At that time, LFTI led the Soviet and international effort in a related area of corpuscular diagnostics of fusion plasmas.[27]

In addition to producing "the brick," a thesis with 100-150 typewritten pages, the candidate also prepared its extended summary, known as "*avtoreferat*" in Russian (literally meaning "auto-synopsis"). The latter had typically 15-20 pages (Fig. 3.9) and listed the official reviewers, the as-

---

25. Named after Igor V. Kurchatov, the science leader of the Soviet atomic bomb program in the 1940s.
26. Gott, 1978
27. Medley et al., 2008; Petrov et al., 2021

3. Blazing My Own Trail   Mike Gruntman

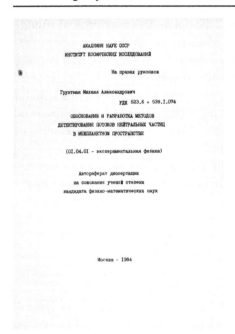

Fig. 3.9. Cover of the synopsis (avtoreferat) of the Ph.D. thesis, 1984, of the author of this book.

signed reviewing organization, and the place and time of the defense, which had to be open to the public unless the work was classified. The avtoreferat then summarized the main points of the defended work, described its conclusions and specific advances of the state of the art, and listed related publications of the author. As required, IKI printed one hundred copies of the avtoreferat in advance and sent them to research physics institutions of the country and main national libraries.

The avtoreferat listed my 16 publications, which reflected the breadth of the Ph.D. thesis. I was the lead author in all except one article. In half of the publications, my sole authorship reflected the situation that I worked largely on my own and without much help.

There was an obvious price to pay for independence and for attempting to tackle a seemingly hopeless experimental challenge of chasing ENAs in space. On the other hand, science life becomes much more interesting and intellectually rewarding when one develops own interests rather than follows tight guidance by the advisor. As Lewis Grizzard used to say in his routines, "Life is like a dogsled team... If you're not the lead dog, the scenery never changes."[28]

The inherent peculiarities and pathologies of the socialist system slowed down the progress of work at IKI. Common sense and realities of life suggested, however, practical ways to mitigate the existing boundary conditions and boost a space science journey to the stars through ... alcohol. Not in a way that the first modern ballistic missile A-4, popularly known as V-2, utilized ethyl alcohol and liquid oxygen for rocket propulsion during World War II, but as an adult beverage.

---

28. e.g., Grizzard, 1995

4. Per Aspera et cum Alcohol ad Astra

Rich in poverty and poor in riches

Many IKI scientists and engineers prepared technical requirements to flight instruments for industrial contractors to design and build. Then they took part in testing the instruments and "mated" them with spacecraft. The latter involved design bureaus of the Ministry of General Machine Building responsible for ballistic missiles and space vehicles. A space experiment thus required the cooperation of IKI specialists, industrial organizations making flight instruments, and engineers who designed spacecraft.

The laboratory of Leonas substantially differed from many units at IKI because it focused on laboratory experiments. Therefore, our life more resembled that of physicists working in other, non-space institutes of the Academy of Sciences. Experimental physics in the USSR combined poverty and inefficient use of rich resources.

We lacked almost everything—and almost always. One had to invest significant effort and time to find components, parts, materials, and vacuum and electronic equipment. Our Western peers could concentrate on science and simply order all those "small things" from catalogs, funds permitting. In addition, much of the domestically produced scientific equipment in the Soviet Union was hopelessly outdated and obsolete.

Moscow periodically hosted international industrial and trade exhibitions in various areas of technology. Foreign governments, particularly the United States, also occasionally held exhibitions in Moscow, highlighting various aspects of national life (Fig. 4.1). Looking through catalogs of equipment and electronics at such exhibitions left no doubt about the general backwardness of Soviet production.

4. Per Aspera et cum Alcohol ad Astra — Mike Gruntman

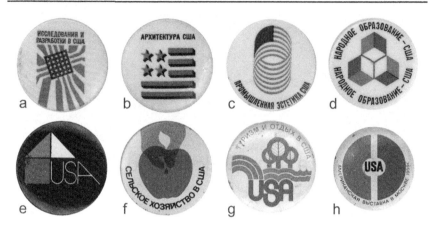

Fig. 4.1. Lapel pins of American national exhibitions in Moscow (primarily in the 1970s and early 1980s). a – Research and Development in the U.S.A.; b – Architecture in the U.S.A.; c – Industrial Esthetics in the U.S.A.; d – Education in the U.S.A.; e – exhibition in an unknown field; f – Agriculture in the U.S.A.; g – Tourism and Vacations (Recreation) in the U.S.A.; h – American [National] Exhibition in Moscow, 1959. From collection of Mike Gruntman.

This backwardness, however, forced thinking and sometimes led, in absence of "brute-force" solutions based on superior modern technology and computational power, to innovative approaches. One could also often spend significant funds on whatever was possible to find in the country without much control and oversight.

In one example, obtaining a better vacuum required baking vacuum chambers at high temperatures. Therefore, we replaced standard rubber gaskets and O-rings with those made of a special synthetic rubber elastomer, Viton, compatible with operations at elevated temperatures. By luck, we found and bought a large number of Viton plane sheets, the only form and shape available. Machinists in the institute's workshop lathed individual gaskets. This was a wasteful and ineffective method but the only one open to us in a centrally planned economy. We purchased perhaps ten times more Viton than needed as nobody knew whether one could find it again. The hoarded Viton could be bartered, in common practice, for something else in the future.

At one international exhibition, I saw metal clamps (Swagelok type), somewhat resembling handcuffs, to quickly connect low-vacuum pipes. Roughing pumps and associated pipes provided preliminary pumping and picked up the exhaust of high-vacuum pumps of experimental chambers.

These clamps truly impressed me by their convenience. However, we could not buy them because the purchase required hard (convertible) currency.

Without any experience, I decided to make my own clamps. I could not obtain the company catalog at the exhibition and saw the clamps only from a distance of ten feet. Therefore, I relied on my understanding of the concept.

IKI's experimental workshop spent hundreds of hours of machinists' time making parts as I experimented with the tightening clamps, tube endings, and gaskets. Finally, the workshop made a large number of working contraptions in my own unique "standard." Our group used these clamps and tube endings for 25 years. They enabled quick and convenient modifications of experimental setups. Consequently, instead of spending an equivalent of a few hundred dollars on commercially available (under normal circumstances) and better designed and fabricated clamps, I used probably one hundred times more resources in addition to my own wasted time.

Small dedicated microcomputers controlled our partially automated physics experiments. We became enthusiasts of the CAMAC system. The abbreviation CAMAC stood for computer-aided measurement and control. Nuclear physics centers and high-energy particle accelerators worldwide widely used the system. It consisted of interchangeable electronic modules with standardized mechanical, electrical, and signal exchange interfaces. Various manufacturers produced the modules in many countries. Consequently, CAMAC allowed physicists to assemble electronic circuitry performing complex real-time data acquisition and computer-controlled functions with minimal development of specialized electronics.

At one time, we got access to CAMAC modules manufactured in Poland. We bought all the modules that we could lay our hands on whether we needed them or not. As a result, dozens of various modules filled the shelves for possible barter.

Universal hard currency

Availability of funds for purchases of equipment, materials, and supplies was important and necessary but certainly not sufficient. Those few "chosen" scientists who were "close to the Sun," that is, those with bosses who were particularly powerful in the Soviet hierarchy, could sometimes use hard currency to buy foreign equipment and science instruments. For everybody else and everyday work, alcohol served the role of a universal hard currency and was widely accepted everywhere in the country.

Researchers in experimental areas typically received certain amounts of concentrated ethyl alcohol, C_2H_5OH, for technical use. Government bodies

4. Per Aspera et cum Alcohol ad Astra   Mike Gruntman

approved allocations of alcohol and strictly controlled its use for such purposes in science and engineering. Ethyl alcohol, also called ethanol, served to clean optical surfaces, electric contacts, and inner walls of high-vacuum chambers and devices therein. In contrast to poisonous methanol, CH_3OH, one could drink pure ethanol, the basis of hard liquors such as vodka or whiskey. Locked safe boxes of experimental groups across the gigantic country guarded the most treasured laboratory possessions, including alcohol.

In the 1980s, my colleague Sasha Kalinin and I received 4.5 liters (1.2 gallons) of pure alcohol monthly for technical support of our high-vacuum experimental setups. One-third of the allocation went for necessary periodic wiping of inner walls of vacuum chambers and washing and cleaning machined parts for use in vacuum. The hygroscopic properties of alcohol facilitated the removal of water adsorbed on the walls, which was essential for attaining and maintaining a clean high vacuum. The remaining alcohol "paid" for purchasing "goods and services" without which no experimental group could effectively function. Very little went to "internal consumption," that is as adult beverages, as neither Sasha nor I abused it.

IKI had a very good experimental workshop with diverse capabilities and many experienced machinists. It could take more than one month to have a certain part made there, following the official route. One began with an engineering drawing of the part. The chief technologist of the institute usually quickly approved the drawing and then the order simply waited in line for its scheduled execution. (I benefited significantly and learned a lot from this knowledgeable technologist at the beginning of my work. Physics education clearly lacked the basics of the technology of machining.)

The everyday work of an experimental physicist required having some small parts lathed and milled. Therefore, each of us had our "own" favorite lathe and mill operators as well as fitters and welders. The worker could do, unofficially, a required simple job within a day or two, or even within a few hours in an emergency. Alcohol played the role of currency to pay for such expedited services. More complex tasks required several hours or even days of effort or needed the participation of different workers. Then an overseeing engineer, a boss in the workshop, managed such off-the-books work and was also getting a "cut" and remunerated by alcohol.

Typically 100 to 150 milliliters (3-5 ounces) of alcohol were paid for a small unofficial machining task. By volume, 200 milliliters (6.5 ounces) of pure alcohol corresponded to a standard half-liter bottle of vodka. (Concentration of pure alcohol for technical needs approached 100%. Vodka and whiskey are 80 proof or about 40% of alcohol by volume). To put it in perspective, for many decades, the daily average salary in the national economy equaled the cost of 1.5-2 bottles of vodka bought in a store. So, 100-150 milliliters of pure alcohol carried a substantial value.

Electricians and technicians in other supporting services of the institute also operated similarly. For example, workers reporting to the chief mechanic of IKI had a foreign-built manually operated hydraulic forklift. We always "rented" this forklift, paying by alcohol, to move a heavy mass spectrometer or a vacuum chamber.

On "friends" and "friendship"

Sometimes in the morning, especially on Mondays, we heard a knock on the door. One of our "friends," a fitter or an electrician, would ask for a little bit of alcohol to cure the heavy weekend hangover. In exchange, he would bring some tools, or wires, or switches, or some other small things always needed in an experimental group.

The IKI leadership knew about such practices, but the demands of producing results forced them to close their eyes. Otherwise, much of the unstoppable progress in reaching the stars would have slowed down.

The supply department of the institute collected annual procurement requests from experimental groups for needed electronics and other equipment. Half a dozen groups could request high-frequency oscilloscopes, for example. The institute would get only one or two, as allocated by the Academy of Sciences. A fair price for receiving this oscilloscope by you and not by another group was one liter of alcohol as a "present" to the head of the supply department. To arrange such an "illegal" exchange, one had to know the person well, to be among his "trusted friends."

Human relations always played an important role at IKI as in any other environment. One could not simply "pay" by alcohol to those who helped in work. It was necessary to maintain friendly relations, gain trust, and show "respect" to the partners, especially when such punishable alcohol-based exchanges clearly broke the rules.

Unfortunately, many IKI machinists liked fishing. I had to politely and patiently listen to unending stories about the fishing accomplishments of my "guys" and occasionally fake interest in their fishing techniques. Going through this torture assured priority of my requests for machining parts.

My colleague Evlanov had excellent connections in other research institutes. He often helped find and "buy" (paying by alcohol) the necessary components and devices for my projects. A few times I returned him a favor helping in transactions for his work.

4. Per Aspera et cum Alcohol ad Astra   Mike Gruntman

One of his friends worked in a major Soviet rocket and space design bureau established by Sergei P. Korolev, today's "Energia." Since I lived nearby in Podlipki, a few times, I picked up bars of lithium metal from Evlanov's friend. It looked like a clandestine operation in a spy movie. The "contact" walked out through the gate and met me behind the nearby trees. We checked that nobody watched us. Then, he transferred the lithium bars hidden under his belt to my briefcase. In a moment, my contact packed back, under the belt, a flask with alcohol.

I quickly established working relations with many machinists at IKI. My colleagues Evlanov, Umansky, and Zubkov (Fig. 2.4) introduced me to the heads of various supporting units. They knew all the right people. I am very grateful for this indispensable education and quick acclimatization in the institute. Without these connections, many developments in position-sensitive detectors and energetic neutral atoms would not have occurred in the 1980s.

After three years of "apprenticeship," I could work in the institute on my own and obtain almost any "goods and services" existing there, either through official channels or by bartering alcohol. I must add that many scientists, especially theoreticians and specialists engaged in space data analysis, as well as engineers who pushed papers with technical requirements and contracting, lived in a different universe. Often naive, they had no clue about this dimension of life in the institute.

As an old Latin saying put it, *per aspera ad astra*, through difficulties to the stars. In our case, indispensable alcohol fueled the progress to space, *per aspera et cum alcohol ad astra*.

Squatters on technical floors

The IKI building had technical floors sandwiched between regular floors with offices and laboratories (Figs. 1.18, 1.19, and 4.2). This space at first remained largely unused. We placed mechanical vacuum pumps there directly under our laboratory and connected them to the high-vacuum chambers through the holes in the floor of the room. It was an excellent arrangement as the noisy mechanical pumps produced exhaust. A small mesh cage surrounded the pumps.

After some time, Kalinin and I accumulated 15 liters (4 gallons) of alcohol for our grandiose construction undertaking. We hired the institute's workers who built a nice large enclosure, about 3 by 8 m (10 by 25 ft), around our pumps with the steel-and-iron floor-to-ceiling walls. The workers welded together metal bars and sheets, built numerous shelves for storage, added electric power lines, installed lighting, and mounted a door with a lock. It became an additional room for storage.

Our expansion took place without any permission or knowledge of the institute administration. The workers themselves found and "procured" or "expropriated" metal sheets and bars and did all construction during their working hours. We were among the very first such "squatters" in the institute. Other groups also began building similar rooms, following the examples. Later, the administration legitimized expansions and filled technical floors with all kinds of properly built rooms as manifested by numerous air-conditioning units seen at their windows today (Fig. 4.2).

Fig. 4.2. The main entrance of the IKI building in June 2016. Large windows are those of offices and laboratories on the high-ceiling main floors. Narrow windows in between are at the low-ceiling technical floors. Visible air-conditioning units at many windows represent the only change in the building appearance from the 1970s and 1980s. This bourgeoise convenience did not reach ordinary folks in the old glorious Soviet days. Photograph courtesy of Mike Gruntman.

Some enterprising heads of supporting units in the institute created "conveniences" for themselves and selected friends, improving "the quality of life." For example, the chief mechanic of IKI, responsible for moving things and for various mechanisms, built a small sauna. Very few people knew about it. Only trusted friends could "rent" the sauna, obviously paying by alcohol, and spent a couple of hours there during a working day.

Vigilant guards on duty and on the take

A special unit of the Ministry of Interior guarded the institute. The young men had sub-officer ranks and served on multi-year contracts. They wore civilian clothes and carried pistols.

At the entrance gate, the guards checked IKI identification cards that served as passes of the employees. (Badges on lanyards did not exist at that time.) The institute staff had strictly specified hours of work. For example, scientists worked from 9:30 a.m. to 6:15 p.m., which included a 45-minute lunch break. The institute required special permissions to come later in the day or leave earlier. One had to obtain a special authorization to stay in the building after 8 p.m. Additional permissions were needed for working after 11 p.m. or entering the institute on Saturdays and Sundays.

Heads of laboratories and departments had the right to enter and leave the institute with handbags. Everybody else could leave a bag or briefcase in a storage room and pick it up leaving the institute. The storage room closed at 8 p.m.

An excruciatingly slow process of obtaining official permissions to take something out from the institute required a lot of effort. To avoid waste of time, we usually asked our laboratory head Leonas to carry through the gate a piece of electronics, such as a pulse counter or power supply, in his bag. Then, we could take it for repairs or perhaps to loan to friends in another institution. One also had to smuggle out alcohol for paying for goods and services outside.

The rules obviously strictly forbade taking alcohol from the institute. Consequently, IKI's welders made excellent flasks out of stainless steel. These polished flat flasks could be tucked under the belt and covered by a shirt. A fair price was on the "volume for volume" basis. That is one paid, for example, 0.4 liters of alcohol for a new shiny 0.4-liter flask.

The flasks allowed easy smuggling of alcohol through the guarded entrance gate, especially during cold seasons when everybody wore a winter coat. The operation turned more challenging in summers when one often had only a shirt or T-shirt and no jacket on hot days.

In the real world, a solution must be found for every challenge for work to be done. And the institute guards were human as well. They had their problems and issues, and their work was stupefyingly monotonous. One of them complained to me that by the end of the day, he just mechanically took the identification card passes, opened them, and brought them to his eyes. He could not see anything because of blurred vision from being dead tired.

Many experimental physicists with access to alcohol had their "own" trusted—that is "bought"—guards. One needed to wait for a day when your "friend" was on duty in a shift at the entrance gate and no other employees or guards were present. Then, one could take almost anything from the institute. It was necessary to tell the trusted friend something plausible that taking out this wrapped package or box would help, for example, advance some important flight project in the institute. The guards knew the irrational bureaucratic system and never challenged such explanations. One paid for this with alcohol. I used the process only a few times for my work and never for personal benefit. Naturally, not everybody exercised such selectivity.

For those without special permission to stay after 8 p.m., the best time for taking things out from the institute through the guarded entrance was from 7:30-7:45 p.m. Most of the employees had left for home by that time. Only those remained who pulled a disproportionately large share of what the institute did. They would leave between 7:50 and 8:00 p.m. So there was a short time window when practically nobody passed through the entrance, offering the best opportunity for taking things out from IKI.

The guys from the guard unit also occasionally stopped by at laboratories of their science "friends," especially if some celebration was taking place after working hours. Then, they could get some alcohol as a present just for being a friend. Some guards grew up in rural areas. Their families had been sending them local delicacies such as homemade bacon, pickles, and sauerkraut, an excellent food to accompany drinks. So mutually beneficial barter occurred as well.

On occasions and without

On occasions of holidays, and sometimes without an occasion, IKI employees had small parties. For example, on the last working day before New Year's Day, many groups bought food and had traditional celebrations. Alcohol was always consumed and celebrations sometimes lasted till late in the evening everywhere in the institute. Heads of laboratories and departments often joined the first hour and then left. The "troops," the scientists, engineers, and technicians, then continued partying. IKI was not exceptional in any way. Similar events took place in other institutes of the Academy of Sciences, industrial research and development organizations, and design bureaus across the country.

Most people drank alcohol diluted with water. A 40% concentration of alcohol made the drink similar to hard liquors in its strength. Some preferred to drink shots of pure alcohol straight, following with water chasers. Mixing alcohol with water is exothermic and releases heat. The mixture thus becomes a little bit warm, which does not make a drink appealing. Note that vodka is served chilled under civilized conditions.

Our vacuum chambers routinely used cold traps to improve vacuum. Consequently, we always had liquid nitrogen around. One could pour liquid nitrogen from a Dewar flask on top of the slightly warm fresh mixture of alcohol with water in a chemical glass beaker. Fumes of evaporating nitrogen, flavored by alcohol smell, would rise. The mixture quickly cooled down to the desired temperature. Sometimes, tiny pieces of ice floated briefly on the surface.

Alcohol certainly helped to better endure such stupidities of a socialist system as assignments to vegetable storage depots. Periodic "temporary duty" at agricultural farms picking up and loading fodder beet (mangelwurzel) and potatoes fell into the same category where alcohol mitigated unhappy circumstances.

We always tried to have the right colleagues included in our detachment to lighten up these wasteful events. One over-naive or irredeemably brainwashed communist idiot full of enthusiasm for fulfilling quotas could make the assignment miserable. With proper organization and foresight, a week-long stint at an agricultural farm turned out much more pleasant with the help of some alcohol and despite the terrible loss of time.

Working hours and after

The most active and productive staff of the institute suffered most from a waste of time in senseless "mandatory-volunteer" activities ordered by Communist Party bodies. The phenomenon permeated the entire country. Not everybody was disappointed with waste, however.

Walking through the IKI building any time during the day one could see groups of scientists, engineers, and administrative staff standing on every floor at designated smoking areas in spacious halls near stairs and elevators. With cigarettes at hand, they passionately discussed the latest movies, soccer games, and their colleagues. Such smoke breaks often lasted for half an hour and were repeated a few times during the day.

As in any organization, the most active IKI staff members worked more than their peers. At the very beginning of my official "life" in the institute, an engineer from a nearby testing department approached me. In a very friendly, caring tone of a mentor and senior colleague (he was in his late forties) he advised me to discuss with my boss how to improve my productivity. Noticing a baffled expression on my face, he continued, "I see that you often stay in the office after official working hours. You obviously do not complete the assignments that your supervisor gives you for the day."

I did not know how to respond. Nobody ever gave me any assignments and I always determined, up to this day, what and how much to do. I always had many more ideas and interests than time to pursue them. Socialism has been effectively killing such attitudes everywhere in the world whenever tried, whether in its soft "democratic" reincarnations or ugly radical forms. It inevitably promoted people according to seniority, connections, and group identity and inherently disregarded merit, celebrating the lowest common denominator. As a columnist of the largest U.S. national newspaper bluntly summarized it, "socialism is for morons."[1] Then, smoke breaks with friendly colleagues looked much more attractive than trying to achieve something tangible in a non-supportive environment.

Most active employees were passionate professionals. They knew each other and helped each other, forming a friendly network based on mutual respect and accomplishment. Sometimes one could hear a knock on a door around 7 p.m. Such a colleague would ask, "Would you like to relax a bit?" This question meant a suggestion to have a couple of drinks before going home around 8 p.m.

Perhaps something happened in his projects and he simply wanted to share a success or maybe seek advice on a technical or administrative issue. Or, the colleague simply wanted to complain about his unfair and not very bright boss. Or, he was dead tired after a long, stressful day.

A telephone could ring with an invitation to join a few friends who gathered in a certain office or laboratory to have *po chut' chut'* (to have a little [drink]). One could have a drink or two or three, accompanied by banter with accomplished colleagues, before going home.

For such cases, I always kept in my safe box a bottle of "flavored" alcohol with orange or lemon peels inside. A good friend and older colleague in our department, engineer Sergei Umansky, sometimes stopped by in the evenings for a chat. The inhumane nature and evil of socialism became common topics of our conversations. Meeting close friends in homes served as another way to maintain a healthy spirit and sanity (Fig. 4.3) in a totalitarian society.

---

1. O'Grady, 2021

Fig. 4.3. The author of this book enjoys a moment in the safety and comfort of a home of friends in the mid-1980s. Relations with close, trusted friends helped navigate the incessant propaganda and pressure of the totalitarian state, especially for those who rejected dehumanizing Marxist ideals and practice.

While the author is with a cigar in this photograph and also shown with a pipe elsewhere in the book, these were truly special occasions. He rarely smoked.

Photograph from collection of Mike Gruntman.

Hole in the wall

Officers maintained the readiness of the institute's armed guards by periodic training in the use of gas masks and at shooting ranges and during political indoctrination sessions. The regime department watched employees and their behavior. The secret department oversaw classified materials. The institute tightly controlled access, required special permission for entry, and limited the visitors to individuals with proper clearances on official business only and representing other similar organizations. A powerful deputy director of the institute, a senior KGB officer, oversaw and coordinated all aspects of security, secrecy, and loyalty of the employees.

This omnipresent system of control, however, missed a hole in the wall—literally, a hole in the wall of the institute.

Perhaps only a dozen people knew about its existence. The hole connected a walled-off area in the basement directly under the low-energy atom-beam laboratory of Evlanov on the ground floor with a storage shack outside the institute's exterior load-bearing wall. The shack housed our large liquid nitrogen tank. It also had some storage space and a door for access from the institute's backyard.

Before my time at IKI, construction workers connected this liquid nitrogen tank to a cooling cryogenic system of a large vacuum chamber in the laboratory hall through a passage in the basement. Incentivized by alcohol, they made the underground opening for pipes bigger than needed. Easily removable foam blocks closed and camouflaged the passage. One could transfer almost anything through this hole. Then, one would go to the backyard, unlock and open the door to the shack, and pick up the "smuggled" piece. It was possible to bring into the institute any item without security people knowing and permitting it. A person could have easily squeezed through that hole in the wall as well.

At first, the backyard of IKI was open, so one could simply walk from our technical shack to a street. Then, in the early 1980s, the institute fenced off the backyard area, with additional entrance gates. These changes diminished the value of our "tunnel."

As in any relations among humans, we had occasional frictions with our guard friends. In such times, my colleague Boris Zubkov liked to fantasize, jokingly, how we could always pay them back. He thought about the dormitories of the *Patrice Lumumba University of Friendship of Peoples* that housed its foreign students not far from IKI. Many of them hailed from developing countries across the world.

The Patrice Lumumba University played an important role in training and indoctrinating pro-communist cadres in professions and building up a Soviet fifth column in Africa, the Middle East, and Latin America. The highest-ranking Soviet-block intelligence official who ever defected to the West noted that to the best of his knowledge, "all foreign students at [the] Lumumba [University] were cooperating, in one way or another, with the foreign branch of the KGB."[2]

A similar university for the same category of students, *Universita 17. listopadu* (The Seventeenth November University), or USL, operated in Prague, Czechoslovakia, from 1961-1974. A cognizant security officer noted that "half the students [at USL] had contacts with Czech intelligence at one time or other."[3] In addition, a school in Leipzig in communist East Germany focused on students from developing countries in another influence operation.

2. Pacepa and Rychlak, 2013, p. 330
3. Frolik, 1976, p. 144

4. Per Aspera et cum Alcohol ad Astra   Mike Gruntman

Some of the "progressive" third-world students in the Lumumba University had a lot of money and liked to patronize nearby watering holes to get drunk. Zubkov used to say that "it would be easy to befriend one such heavily drunken student and bring him into the institute through our hole in the wall. He would never remember how he got there and start wandering through the institute building at night. Then let our guard friends explain to superiors how this student, looking very different from Muscovites, ended inside a heavily guarded facility." This dream plan was never put into action.

Mystery of welded windows

Despite all the layers of control and various administrative bodies, supported by a large omnipresent Communist Party organization with many loyal enthusiastic members, certain events remained mysteries and never uncovered. This happened, for example, with an incident at the "winter garden."

Institute director Sagdeev succeeded in establishing large-scale cooperation with foreign scientists in space research. The institute built a so-called display center on the second floor of one of the building sections. There, foreign scientists and their Soviet counterparts discussed joint space experiments and received data from spacecraft. This was also a comfortable place for science seminars.

Soon, the center expanded with a greenhouse next to it, protruding from the building on the backyard side. Flowers and small trees grew in this winter garden, a remarkable place in cold, dark winters.

One evening, an empty bottle flew out from a window of one of the upper floors and crashed through the glass ceiling of the garden. Fortunately, nobody was there to get hurt at that late hour.

The administration, Communist Party activists, and security units mobilized all their resources, certainly including the informers, to find the culprit and others who took part in the drinking "party." Without success.

Within two weeks, the workers permanently welded shut all the windows on all floors in the entire section of the IKI building above the winter garden. Nobody would be able to open a window anymore. Two months later, the main "hero" of this event pointed at the "culprit" window, now welded shut, and told me, "Well, it happened here, the celebration went just a bit overboard." The institute kept its secrets.

In some cases, the administration knew the perpetrators but chose not to punish them. In one case, a not very sober senior research fellow from a friendly laboratory had an argument with security guards at the entrance gate. Upset, he went back to his experimental hall on the ground floor and left the IKI building through a window. The administration mildly admonished him. A joke went at that time, "Actually, it is a good sign that he got drunk. That means that he is a loyal Soviet citizen and would never betray the socialist motherland [as intellectual dissidents less focused on drinking might do]."

With all these peculiarities of life, our development of position-sensitive detectors and advancement of ENA physics and instrumentation were progressing.

5. Position-Sensitive Detector: First Image and Beyond

Microchannel plates

In the Soviet Union, the *Gran* plant (today's *Baspik*) in the town of Ordzhonikidze in the North Caucasus manufactured MCPs. (The town reverted to its prerevolutionary name Vladikavkaz in 1990.) The plant has been making electro-optical image converters and intensifiers since 1964.[1]

In 1976, we obtained several defective 25-mm (1 in.) in diameter MCPs with a promise of operational plates in the future. Each plate had one million straight miniature channels stacked in parallel (Fig. 3.3) and required that 1,000 volts be applied across it between the entrance and exit surfaces. The plates had to operate in a vacuum better than 10^{-5} tor to avoid damaging discharges.

A single MCP "multiplied" an electron produced by an impact of a particle or photon at its entrance into an avalanche of 1,000-10,000 electrons at the exit. As was found in 1972,[2] one had to use two or three consecutively mounted plates to achieve the desired electron multiplication for the detection of individual particles. Based on journal publications, nobody in the Soviet physics community had experience with such MCP detectors and could advise how to build the stacks of plates. Hence, I began to design my own MCP assemblies, with the workshop making parts from stainless steel compatible with operations in a clean vacuum.

Defective MCPs helped in the development of practical designs. A few times my screwdriver slipped from a slot and went through a pair of such plates during the assembly of detector prototypes. (The convenient Phillips-head screws and the socket screws with hexagonal drive holes could have

1. Ponomarenko and Filachev, 2007, pp. 64, 135
2. Colson et al., 1973

5. PSD: First Image and Beyond   Mike Gruntman

helped, but they were not available to us. I am not sure whether they even existed in the Soviet Union since I never saw them.) The plates turned into useless fragments of black glass. At that time, the price of three operational MCPs equaled my annual salary. Therefore, experimenting with designs and learning assembly skills on real but defective plates was wise.

Soon, a practical design emerged for stacks of two and three MCPs. My assemblies allowed mounting the plates with various widths of the gaps between them and applying independent voltages to all contacts, including across the gaps. These features would lead later to achieving record amplitude resolutions, important in some applications. Many groups in the world mounted the plates directly surface-to-surface, one to another or separating them by a thin conducting ring near the plate edges. Literature would refer to such arrangements of two and three plates as the chevron and Z stacks, respectively.

In a vacuum chamber, we explored characteristics of the detectors registering collimated beams of monoenergetic ions or neutral atoms and under extreme ultraviolet and X-ray photon illumination. As in earlier studies of the VEU-6 channel electron multipliers, our work led to the first publications in Soviet physics journals on the properties of MCP detectors counting individual particles.[3] A very few earlier publications originated from the developers of MCPs in the industry.[4] They typically focused on plate properties and comparison with Western analogs when detecting electrons.

Our existing laboratory setup allowed experiments with ion and neutral particle beams in a relatively small vacuum chamber one meter (3.3 ft) long. I designed and built two large new sections of the chamber to expand the particle path by an extra three meters (10 ft) in a vacuum. This major augmentation would enable future studies of scattering of atoms and molecules excited in collisions by time-of-flight techniques. Leonas advocated this strategic development of experiments. Kalinin and Morozov completed the modification of the experimental setup in the early 1990s.

I had been continuously developing various MCP detectors from 1976 until 1990, with my colleagues Kalinin (initially) and Kozochkina (later) taking part in some work. The studies of characteristics of MCP assemblies led to a detailed understanding of the physics of their operation and optimization of the designs, described in several articles in *Pribory i Tekhnika Eksperimenta*.

Our MCP detectors had achieved record amplitude resolutions by the mid-1980s. This feature enabled measurements of the statistics of electron

3. Gruntman and Kalinin, 1979, 1980
4. Bragin et al., 1975

emission from thin (15-25 atomic layers) foils caused by energetic particles, such as ENAs, passing through them. Exploration of the emission statistics was among my priorities (see 1979 List in Chapter 7) since the late 1970s. This phenomenon provided the foundation for pulse amplitude discrimination in space sensors. Such a technique could assist in noise suppression when detecting hydrogen ENAs[5] and in atom identification in ENA imaging of planetary magnetospheres.[6] We published the results of measurements of electron emission statistics in *JETP Letters*.[7] This physics journal was an analog of the *Physical Review Letters* in the United States and was translated cover-to-cover to English. To the best of my knowledge, nobody has ever repeated such measurements.

Elektron and SNIIP

A few leading physics groups in the West, including at AMOLF, worked on MCP-based position-sensitive detectors for differential scattering experiments. Leonas obtained support from IKI for the development of a PSD for unspecified applications in space. Nobody in the institute showed much interest in such sensors beyond a perfunctory "it would be nice to have." Therefore, we could develop PSDs for differential scattering experiments, as my boss strived for, while the institute paid for it and justified the work by its promise for future space plasma and astrophysics instruments.

At the end of the 1970s, we began the formal development of a space-qualified (flight-qualified) MCP-based PSD with two partners from industrial research and development organizations. Mikhail R. Ainbund led design of an MCP stack with a specialized anode in *Vsesoyuznyi Nauchno-Issledovatel'skii Institut "Elektron"* (*All-Union Scientific Research Institute "Elektron"*), or *VNII Elektron*, in Leningrad. Two laboratories in S*oyuznyi Nauchno-Issledovatel'skii Institut Priborostroeniya* (*Union Scientific Research Institute of Instrument Building*), or *SNIIP*, under Boris I. Khazanov and Lev S. Gorn worked on electronics in Moscow.

From the outside, this program looked typical for IKI. The institute contracted Electron and SNIIP to develop a detector per our technical requirements. Both Electron and SNIIP had experience in building space flight instruments for scientific and defense applications.

In an unusual twist, we were one of the IKI groups engaged primarily in laboratory physics experiments, including the related detector physics.

---

5. Gruntman and Leonas, 1983
6. Gruntman and Leonas, 1986
7. Gruntman et al., 1990a; also, Gruntman et al., 1989

5. PSD: First Image and Beyond

Consequently, our role in the development of the position-sensitive detector and later ENA sensors became unusually large. I had been designing and building my own MCP stacks and studying their characteristics for a few years by that time, which led to a detailed understanding of their physics.

A combination of this advanced expertise in detectors with our experimental capabilities for various testing under controlled conditions of a physics laboratory allowed us to identify problems in prototypes developed by our Elektron and SNIIP partners. Besides, I never shied away from offering—unsolicited—my own ideas and design solutions. Our accomplished collaborators were not accustomed, however, to competition, critic, and suggestions by "outsiders." Consequently, my relations showed some strain with one partner and evolved into a truly joint work with another.

Elektron's Mikhail Ainbund was among the most experienced engineers involved in the technology of secondary electron multipliers. He had participated in the development of the commercially available multidynode electron multiplier VEU-1 (Fig. 3.1) and channel electron multipliers, including VEU-6 (Fig. 3.4). Then, in the 1980s, he led work in VNII Elektron on a stack of two MCPs (without an anode). The Gran plant in Ordzhonikidze began series production of this detector, VEU-7, with a convenient mount and four high-voltage connectors (Fig. 5.1). The detector became popular in physics laboratories in the Soviet Union and worked well in the basic counting of particles and energetic photons.

Fig. 5.1. Commercially available two-MCP detector VEU-7. Photograph courtesy of Yu.V. Gott.

My relations with Ainbund always showed a bit of tension. He and I had different priorities and different internal environments in our home organizations. The engineering culture in the industry certainly differed from that of physicists in the Academy of Sciences. I was also significantly younger but never let it play a role in our scientific and engineering interactions.

Obviously, we could never compete with Ainbund, who relied on the extensive capabilities of a large industrial research and development establishment in the powerful electronics industry. I could only dream of the technologies at his disposal and always respected him personally and for his work. But we also occasionally disagreed. One of my articles in PTE on the single-electron amplitude resolution of MCPs[8] apparently upset Elektron management, as my findings undermined some technical claims important to the officialdom in their ministry.

SNIIP's Boris Khazanov and Lev Gorn designed front-end electronics for our position-sensitive detector. They had been working with IKI for many years, building flight instruments for two rival space plasma groups of Oleg Vaisberg and Konstantin Gringauz. SNIIP specialized in nuclear instrumentation and belonged to the *Ministry of Middle Machine Building.*[9] This ministry oversaw the development and production of nuclear weapons and all other nuclear-related activities in the country, including nuclear power plants, uranium mining and enrichment, and high-energy particle accelerators with associated fundamental and applied physics.

As was typical for the USSR, state antisemitism blocked promotion to higher technical leadership positions for Jews Khazanov and Gorn. They also avoided joining the Communist Party. Consequently, Khazanov and Gorn headed two friendly laboratories in SNIIP and concentrated on their engineering work. Every three-four years, they published a new book on nuclear electronics in Atomizdat, the publishing house of the nuclear establishment.

Very early in our joint work, I went to pick up from SNIIP an electronic unit with charge-sensitive amplifiers, a simplified prototype of the front-end detector circuitry. Rather than going through the hassle of paperwork for getting me a pass to enter the building, Khazanov, Gorn, and I met at a waiting area near the visitor gate. This was a common time-saving practice across the country. They gave me a box with labeled input and output connectors, which was supposed to be a true "black box" for me, and explained what it did and how to use it.

Highly accomplished and respected senior specialists, Khazanov and Gorn were revered and "feared" at IKI. This did not prevent me from telling the sages on the spot that the functions of the box looked wrong to me and it would not work as needed. Very tactfully, smiling, the two wise men explained to me, a young "greenhorn," in unambiguous terms that they knew what they were doing. Intimidated, I could not find the right arguments,

8. Gruntman, 1985.
9. Gruntman, 2015a, p. 96

took the electronic box, and headed in dismay to the nearest subway station.

When I got to the station, I stopped, turned around, and went briskly back. From the visitor gate, I called Gorn on the internal telephone and he came down to see me again. I repeated my concerns. After chatting with Khazanov over the internal telephone, Gorn agreed to take the box back in order "to check my arguments and dispense my doubts." Although he sounded as if I was still wrong, it was clear that they now felt uncertain.

I returned to IKI convinced that I was right. And it turned out to be so. SNIIP corrected the mistake and the new electronics box worked properly. From that moment on, Khazanov and Gorn had "noticed" me. A truly joint work, as partners, followed. I continuously used my own detectors and laboratory data acquisition system to test and optimize their analog electronics, with our expertise being complementary. This interaction differed from their typical relations with other groups at IKI.

Despite the tremendous difference in our status and age, Khazanov, Gorn, and I quickly developed mutual sympathy and friendly relations. After some time, they began sharing with me stories, going back for years, about their projects of building flight instruments for IKI groups. As I always exercised discretion, they told me interesting details about leading lights at IKI, always with humor peppered with sarcasm.

As an example, on one of their visits to IKI, a very distinguished institute researcher and laboratory head (Gringauz) in a fatherly patronizing manner advised Khazanov and Gorn to stop working with another laboratory at IKI (headed by Vaisberg) who did not know, in the words of Gringauz, what they were doing. Then, two hours later, that other distinguished laboratory head (Vaisberg) asked them whether they continued wasting, in his view, precious government resources on developing instrument prototypes for the first laboratory head, the prototypes that, in Vaisberg's words, nobody needed and that would not work anyway.

First image

A typical microchannel plate detector consists of an assembly of two or three MCPs with straight channels, followed by a collector (anode) for capturing the electrons (Fig. 5.2). The stack converts each registered incident particle (energetic electron, ion, ENA, EUV or X-ray photon) into an avalanche burst of tens of millions of electrons that falls on the collector.

Some detectors (such as the one in Fig. 3.3, right) register precisely, with a nanosecond accuracy,[10] the moment of the particle detection. This feature

---

10. One nanosecond is 10^{-9} of one second, 1 ns = 10^{-9} s.

is particularly important for time-of-flight instruments. In position-sensitive applications, the collector signal readout electronics determines the location of the electron avalanche and thus the impact position of the incoming particle on the sensitive surface of the detector.

Our position-sensitive detector included a stack of two microchannel plates followed by a collector consisting of 19 disks (Fig. 5.3). The electron avalanche spreads out laterally in the gap between the MCP exit and collector.

By measuring fractions of the total charge

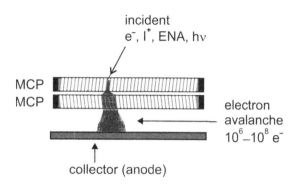

Fig. 5.2. Microchannel plate detector with two plates in sequence with sightly tilted straight channels in the so-called chevron configuration followed by a collector (anode). In position-sensitive detectors, the collector signal readout system determines in real time the position of the center of the electron avalanche (10^6–10^8 electrons) created by each detected particle. This avalanche position unambiguously corresponds to the coordinates of a point where the particle or photon that triggered the avalanche hits the sensitive surface of the front MCP.

intercepted by the disks, one could determine the position of the avalanche center. This readout concept resembled an approach implemented in a

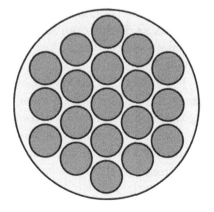

Fig. 5.3. Collector of electrons, an anode, consisting of 19 disks in a hexagonal pattern for IKI's MCP-based position-sensitive detector.

5. PSD: First Image and Beyond   Mike Gruntman

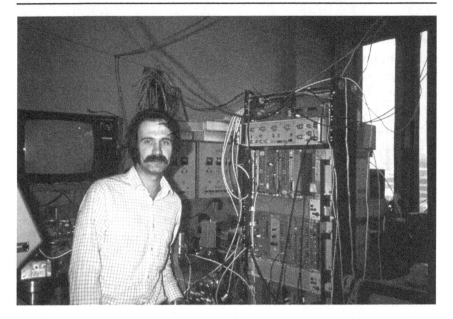

Fig. 5.4. The author of this book in his laboratory-cum-office in the IKI building (room 419) in the mid-1980s. An electronics rack on the right shows two CAMAC crates with various modules; a pulse counter is on the top. In the back, one can see control panels for turbomolecular vacuum pumps on the wall on the right and a computer-controlled television set for image display on the left. Photograph from collection of Mike Gruntman.

scintillation camera for imaging human organs *in vivo* in gamma rays by Hal Anger in the late 1950s.[11]

In early 1982 I assembled the entire detector system in my vacuum chamber. Ainbund built a spaceflight prototype MCP assembly with a 19-disk collector. The detector sensitive area was 24 mm (0.94 in.) in diameter. A 10-mm gap between the last MCP exit and the collector had a uniform axial electric field.

The groups of Khazanov and Gorn built the front-end electronics consisting of 19 charge-sensitive amplifiers followed by an analog circuitry determining the coordinates of the centroid of the electron avalanche hitting the collectors. I designed and built a system, based on the microcomputer-controlled CAMAC modules, that picked up analog pulses at the outputs of the front-end electronic unit, digitized their amplitudes, and displayed positions of the registered individual particles on a TV monitor (Fig. 5.4).

---

11. Anger, 1958

Fig. 5.5. Standard electronics supporting physics laboratories in the 1970s and 1980s: oscilloscopes, voltmeters, pulse counters, power supplies, and others. A typical drawing board with a drafting machine stands in the back. Photograph (in IKI museum, 2018) courtesy of Mike Gruntman.

The relatively slow CAMAC system could register and determine coordinates of 100 particles (photons) per second in real time. A microcomputer controlled the operation of the laboratory prototype of the detector with a code in a high-level interpreter language. The optimized flight instrument with custom-built electronics could have processed and registered up to 50,000-100,000 particles per second. Other common electronic equipment such as oscilloscopes, counters, power supplies, voltmeters, and others supported the work (Fig. 5.5).

Ion gauges measured background gas pressure in vacuum chambers. Turning on such a gauge created ions that were attracted and accelerated to 2 keV energies by the negative voltage applied to the input surface of the front MCP. These ions thus bombarded the sensitive area of the position-sensitive detector. In front of the detector, I installed an easily recognizable fragment of a familiar double-edge razor blade (Fig. 5.6, top).

The assembled system sequentially determined the X and Y coordinates of impact points on the sensitive area of each registered ion, stored them in the memory of the computer, and lighted up the corresponding pixel on a TV

5. PSD: First Image and Beyond   Mike Gruntman

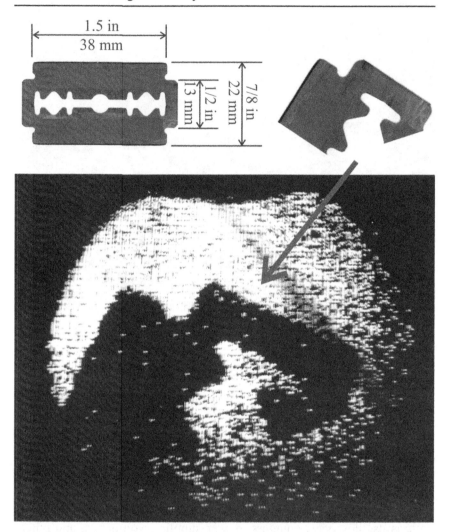

Fig. 5.6. Photograph of the display screen with the first recorded image of a shadow cast by a piece of a standard double-edge razor blade formed by incident ions on a sensitive surface of the MCP-based position-sensitive detector (bottom). A couple dozen white dots (pixels) across the shadow were the detector intrinsic noise counts. This was the first image obtained by a PSD behind the Iron Curtain in 1982.

A razor blade with its dimensions is shown on the top left. On the right is a fragment of the blade installed in front of the detector.

The PSD system sequentially registered incoming individual ions, determined their X and Y coordinates, and lighted up the corresponding pixels on the screen.

Photograph courtesy of Mike Gruntman.

monitor. After a few minutes of accumulation of the signal, one could clearly see on the screen a shadow cast by the blade fragment (Fig. 5.6, bottom). The first MCP-based PSD in the Soviet Union and the entire Eastern Block behind the Iron Curtain produced its first image.

I obtained this very first image around 3 p.m. and showed it to Leonas. My boss immediately went to a deputy director of IKI for instrumentation, Vyacheslav M. Balebanov, to bring him in and share this breakthrough into truly enabling instrumentation. This detector technology could have dramatically advanced the capabilities of many plasma and astrophysics space instruments of the institute. Balebanov did not have time then and stopped by a few weeks later. Leaders of various research groups did not show much interest at all. At least, they did not express it publicly.

In January 1983, we submitted a joint article (IKI, VNII Elektron, SNIIP) to *Pribory i Tekhnika Eksperimenta* describing the detector. The article appeared in print in early 1984.[12] (It typically took 12-18 months to publish an article in PTE at that time, on top of the additional three months for obtaining internal permissions at IKI and the Academy of Sciences to send a manuscript to a journal.) The same issue of the journal also featured my review article[13] on MCP-based position-sensitive detectors, perhaps the first comprehensive review article on such detectors in a mainstream physics journal in the world.

Unfortunately, I lost a battle with my boss on terminology. I wanted to call the detectors *position-sensitive* in direct translation into Russian, a linguistic calque, of the terminology accepted in the English-language literature worldwide. Leonas, however, insisted on the Russian term *coordinate-sensitive detectors* and overruled me. As a result, Western compilations of scientific publications often did not properly translate and classify our articles and they lost some exposure. I could only partially mitigate this "lost-in-translation" problem by directly communicating with colleagues abroad.

Wedge-and-strip collector

During our work on the 19-disk collector PSD, a new important publication appeared in the journal *Review of Scientific Instruments* in July 1981. Scientists from the Space Sciences Laboratory (SSL) at the University of California in Berkeley described an elegant design of a position-sensitive detector with a collector consisting of only three elements.[14] The article

---

12. Ainbund et al., 1984
13. Gruntman, 1984
14. Martin et al., 1981

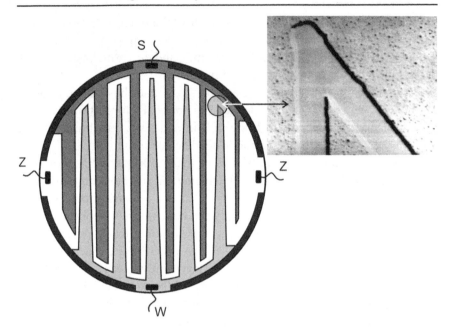

Fig. 5.7. Left: simplified conceptual schematic of a wedge-and-strip collector of electrons. It consists of three electrodes only: identical electrically connected wedges (light gray, W); electrically connected strips of varying width (dark gray, S); and the "snake" electrode in between (white, Z). Right: electron microscope photograph of a 16-micron wide insulation gap between two electrodes of a wedge-and-strip collector fabricated by photolithography (Gruntman et al., 1990b). Photograph from collection of Mike Gruntman.

authors included Anger, of gamma-ray detector fame, and Mike Lampton, with whom I would later establish excellent collegial relations. The complex geometry of this so-called wedge-and-strip collector required precise photolithography for its fabrication (Fig. 5.7). On the positive side, it significantly simplified electronics.

The *Space Astrophysics Group* in SSL built the wedge-and-strip PSD. Stuart Bowyer headed this group, which focused primarily on space-based extreme ultraviolet astronomy. They had an outstanding record of innovation in various detectors and instrumentation. The T-shirt of the group prominently displayed a bold, with chutzpah, motto: *Veritas Duo Sigma*. This Latin phrase meant *two sigmas make truth*, that is, two standard deviations instead of the usual conservative and cautious three.

The journal *Review of Scientific Instruments*, where the article appeared, was and remains one of the most prestigious publications in the world on general techniques of physics experiments. The Soviet Union translated (until July 1991) this journal cover-to-cover into Russian and widely distributed it among research institutions. As mentioned in Chapter 3, Plenum translated, in turn, cover-to-cover a Soviet analog of the journal, *Pribory i Tekhnika Eksperimenta*, into English. The Institute of Physics in the United Kingdom published the third similar leading general physics instrumentation journal, *Journal of Physics E*. It was for an article in the latter that I received a lecture on "the patriotic duty of a Soviet scientist" from a security official at IKI (Chapter 2).

The IKI library received the original English version of *Review of Scientific Instruments* but always with a delay of 6-8 months for some unknown reasons. So, Leonas and I saw the article about the wedge-and-strip position-sensitive detector in early 1982.

Huge inertia characterized the centrally-planned Soviet system. Consequently, we could not modify the ongoing contractual development of our position-sensitive detector with the slowly-moving industrial partners. We could, however, try to build this new promising detector on our own. Therefore, the joint IKI-Elektron-SNIIP work on the 19-disk collector PSD continued unaffected, and soon I obtained the first images described above.

At the same time, I rushed to reproduce the wedge-and-strip detector. Leonas found connections in the semiconductor industry to make the collector. The photolithographic process required first to produce a precise enlarged drawing of the collector, about one meter (3 ft) in size. The drawing was then reduced to make a photomask with micron precision (Fig. 5.7, right) that was used to fabricate the collector 4-5 centimeters in size.

IKI had one foreign-made wide-format printer (Benson) supporting a mainframe computer. This printer could plot drawings of the desired dimensions but it was very hard to get permission to use it. It happened, however, that among my good friends were engineers from computer units. They immediately turned the green light on for me without knowledge of their supervisors, accompanying the arrangement with clinks and toasts for success. It took me one month to design the collector, write a code for the mainframe computer, and plot the required drawing. In the process, I used up a lot of imported, and thus strictly controlled, special paper and ink.

One more month later we had a few fabricated wedge-and-strip collectors in our hands. The industry made them without any formal contracts and purchase orders. Leonas took two liters of alcohol to his friend in a secret semiconductor establishment who pushed through our "special order" in their photolithographic units. People in the defense industry did not differ

5. PSD: First Image and Beyond — Mike Gruntman

much from those in an academic IKI, and they readily accepted the universal hard currency of alcohol for a good cause.

Using existing electronics, I quickly built a CAMAC-based signal processing system and assembled the new wedge-and-strip collector and one of my MCP stacks. For the first and only time in my life, I also designed and built a relatively simple electronic CAMAC module to control the acquisition of detector signals. It was fun.

Soon, our wedge-and-strip PSD demonstrated its imaging capabilities. Overall, it took four months to obtain the first images from the moment that Leonas and I had seen the Berkeley article in *Review of Scientific Instruments*.

The first images turned out, however, to be spatially compressed (smaller in size). After one week I pinned down the cause, a large capacitive coupling among the elements of the wedge-and-strip collector. The fix required properly optimized amplifiers. I kept SNIIP's Khazanov and Gorn fully informed about our progress as they were also very much interested in this new detector concept. Very quickly, they provided new charge-sensitive amplifiers for me, without any formal contracts. This solved the problem.

Position-sensitive connections

The Space Astrophysics Group in Berkeley's Space Sciences Laboratory stood out with its many achievements in MCP-based position-sensitive detectors. In 1983, the Academy of Sciences authorized us to invite one of its leading specialists, Mike Lampton. Mike and I had begun exchanging detector publications by mail a couple of years earlier. The Academy also invited Mike's wife Susan, a university physics professor. In addition to IKI, their trip included a visit to LFTI in Leningrad and a space science group in the Georgian Academy of Sciences in Tbilisi.

A recognized authority on applications of microchannel plates,[15] Lampton contributed to several breakthroughs in position-sensitive detectors, including distortionless resistive anodes and wedge-and-strip collectors. Later, he advanced readout techniques based on delay lines. In addition to Mike's other consequential accomplishments in optics and EUV astronomy, NASA selected him for the astronaut corps (Fig. 5.8). He underwent training to fly on the Space Shuttle as a payload specialist and would serve on a backup crew of the first flight of the Shuttle's Spacelab in a mere half a year after the visit to Moscow.

Meeting Mike Lampton in person provided a unique opportunity to

---

15. e.g., Lampton, 1981

Fig. 5.8. Mike Lampton, Space Sciences Laboratory, University of California, Berkeley, in an official NASA astronaut photo in the early 1980s. Photograph courtesy of NASA.

validate my work on MCP detectors. Our understanding of the detector physics was really good, but we obviously could not compete with Berkeley because of our country's general backwardness in components and materials. Working largely alone at IKI did not help either.

An outgoing man of many talents, Lampton was fourteen years older than I. We quickly formed a friendship that lasts to this day. The interaction with Lampton proved to be particularly useful because I could not travel beyond the confines of the tightly guarded socialist camp. For several years, the stream of publications mailed from Lampton's group helped me stay abreast of the state of the art in position-sensitive detectors in the West. It was also Mike who soon connected us for the first time with scientists in the *Applied Physics Laboratory* in the United States, who had begun exploring energetic neutral atoms in planetary magnetospheres. Personal communications always played an essential role in science and helped defeat irrational restrictions.

One day in the mid-1980s, a French space scientist stopped by our laboratory during his visit to IKI. After listening to my description of the work on MCPs and PSDs, he called me "Mike Lampton of the Soviet Union." It was flattering and made me feel proud.

Our advances in MCP-based position-sensitive detectors in the early 1980s brought recognition among colleagues in the Soviet Union and got noticed abroad.[16] It could have opened wide opportunities to expand successful work in a "normal" common-sense environment. But it did not happen.

---

16. e.g., Timothy, 1984, 1985

5. PSD: First Image and Beyond   Mike Gruntman

No road to independence

For several years I had been in a unique position as probably the only physicist in the Soviet Union who could realistically build MCP-based position-sensitive detectors. Many groups in research institutions would have benefitted from using such detectors if they were available. Making them, however, was truly non-trivial and required highly specialized expertise in various fields. A new commercially available stack of two MCPs, VEU-7 (Fig. 5.1), allowed counting various particles but lacked position-sensing capabilities.

A small dedicated group was needed to make position-sensitive detectors while I worked practically alone in the institute. The industry could have obviously done it. However, the centrally planned socialist behemoth required high-level government decisions and cumbersome cooperation of various slow-moving organizations from different "stovepipes," reporting to different ministries.

Arranging such an undertaking did not interest me and looked unrealistic. It would have taken many years, with the technology becoming obsolete in the meantime. In addition, it could have only been done by somebody with a Communist Party membership card and comfortable in being an inalienable part of the Soviet political system.[17] I was not in that category and did not want to become part of the regime. The attitudes of my bosses in the institute presented another obstacle.

I was convinced in those years that it was possible to make position-sensitive detectors, including spaceflight-qualified sensors, at IKI for various research groups in the institute and other outside organizations. Such an endeavor could have only happened if IKI had allowed me to grow. This required the creation of a small unit within the institute, perhaps half a dozen scientists, engineers, and technicians, operating as a semi-independent small company. My expertise, enthusiasm, and energy sufficed, but neither the institute nor my immediate bosses in the department supported me in this desire. The latter obviously did not want to lose me as a subordinate. The upper levels of IKI simply did not care. The road to independence and growth was blocked.

In the United States, at least one small company established itself in California in this particular niche, building MCP-based position-sensitive detectors, in the 1980s. Then, another company emerged in the early 1990s.

---

17. Under a similar environment, the Polish language developed an appropriate word, the adjective *reżymowy*, describing somebody who was inherently part of the regime.

Something similar could not happen and has not happened in the Soviet Union. The economic system and the culture of the society—in short, socialism—were inherently incapable of advancing such innovation and encouraging independence.

Glacially slow advance

It became clear with time that my ambitious activities in position-sensitive detectors would never be allowed independence and expansion at IKI. Under real-life constraints of the department and institute, only a slow advance could happen and only through external cooperation with others interested in such detectors or their elements. Our early accomplishments opened many doors for such collaborations.

Position-sensitive detectors would have benefited from larger sensitive areas. So, in the second half of the 1980s, we designed and built such large MCP stacks in a joint effort with a group headed by Boris M. Glukhovskoi at the *Moscow Electronic Valve Plant*.[18] The predecessor of this organization, Plant No. 632, produced electronic image converters as early as 1942. It had been building converters and image intensifiers since then and possessed the expertise and technologies needed for working with MCPs.[19] A former Fiztekh student Alla Kozochkina, who had become a staff member in Department No. 18, participated in this work. A university group in Nalchik in the North Caucuses also teamed up with us to make a similar detector for their laboratory physics experiments.

The year 1985 brought consequential interest and support from an unexpected direction. Victor Afanasiev, a new director (from 1985-1993) of the *Special Astrophysical Observatory* (SAO) of the USSR Academy of Sciences, got interested in photon-counting imaging detectors in the visible spectral range. Energetic Afanasiev represented a pleasant change from typical slow-moving science managers in the Academy of Sciences. He found out about our detector work at IKI and one day followed up with a telephone call asking me to urgently fly to an astronomical workshop in Kiev (Kyiv) to meet him and his associates.

SAO operated a 6-meter (19.7-ft) in diameter optical telescope (Fig. 5.9). It was located in picturesque mountains at an altitude of 2070 m (6800 ft) on the northern slope of the western part of the Greater Caucasus Mountains. The *Large Altazimuth Telescope* (*Bol'shoi Teleskop Azimutal'nyi*, or, *Al't-Azimutal'nyi*, *BTA* in Russian), BTA-6, was the biggest in the world

---
18. Blatov et al., 1990
19. Ponomarenko and Filachev, 2007, pp. 64, 135

5. PSD: First Image and Beyond — Mike Gruntman

Fig. 5.9. Dome of the Special Astrophysical Observatory's Large Altazimuth Telescope BTA-6 with a 6-m (19.7-ft) mirror in the 1980s. The width of the open slit (left) exceeded 11 m (36 ft) and the rotating cupola weighed 1,000 metric tons. BTA-6 was the largest optical telescope in the world from the mid-1970s until the early 1990s. Photographs from collection of Mike Gruntman.

Fig. 5.10. Commemorative envelope (left) and postage stamp (right) with the 6-m telescope, the largest in the world. The envelope and stamp are not to scale. From collection of Mike Gruntman.

at that time. The number 6 in its name stood for the telescope diameter in meters.

The observatory was a highly visible organization and gave the country bragging rights. The government supported it with ample resources. Specially issued collectible envelopes and postage stamps celebrated this achievement of Soviet astronomy (Fig. 5.10). After an initial period of difficulties, the telescope became fully operational, although the astroclimate (astronomical seeing) at the location was far from ideal. A small hotel next to the dome at the mountain top (Fig. 5.11) provided accommodations for visiting astronomers.

A supporting settlement, Nizhnii Arkhyz, 800 m (2600 ft) lower, housed the scientists, engineers, and staff of SAO (Fig. 5.12). The Special Astrophysical Observatory also operated a radio telescope, RATAN-600 or *Radioastronomicheskii Teleskop Akademii Nauk* (*Radioastronomical Telescope of the Academy of Sciences*). Almost 900 rectangular reflectors formed the telescope collecting area, a circle 600 meters in diameter (Figs. 5.13, 5.14). RATAN-600 was located near the settlement Zelenchukskaya, 12 miles (20 km) away from Nizhnii Arkhyz and 1100 m (3600 ft) lower in altitude.

A highly sensitive photon-counting imaging detector in the focal plane of SAO's BTA-6 would have opened new interesting observational possibilities. Operating MCP-based PSDs in the visible range required sealed detectors with photocathodes and thus the involvement in their development of additional partners with highly specialized expertise. Making detectors with proximity photocathodes in vacuum-tight sealed enclosures with transparent windows relied on particularly clean materials and advanced vacuum technology. Therefore, such an imager in the visible range presented even bigger technical challenges than our already complex open-type (that is, operating in vacuum) detectors.

SAO director Afanasiev got interested in position-sensitive detectors when an astronomer from Mexico city, Claudio Firmani, suggested bringing his photon-counting imager Mepsicron[20] for observations on the 6-meter telescope. It was dark and cold in December or January when Firmani came to SAO on an exploratory visit. I flew there as well to meet him. A group in Berkeley, including Lampton (small world!), made Firmani's Mepsicron, so it was a high-quality state-of-the-art sensor. Afanasiev felt that we had the resources to build such a detector as well.

SAO quickly secured funds to launch a joint program to make photon-counting imagers for the focal plane of the 6-meter telescope. A specialized

---
20. Firmani et al., 1982

5. PSD: First Image and Beyond   Mike Gruntman

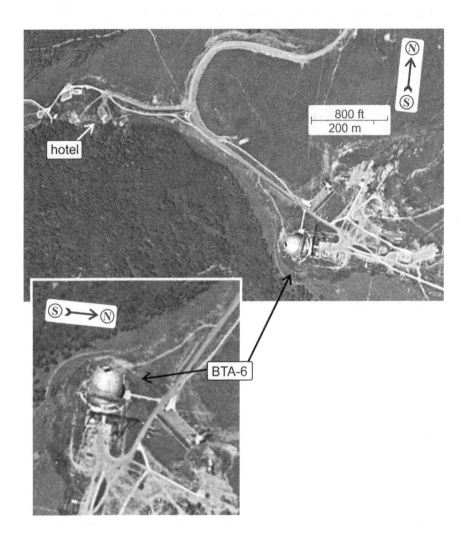

Fig. 5.11. Satellite photograph of a mountain top with BTA-6 (43°38.8' N, 41°26.4' E) and a small hotel for visiting astronomers nearby (top). The insert (bottom left) shows the magnified fragment with the telescope dome and a metal gantry in front (also seen in Fig. 5.9, right) for handling large telescope components. The photograph is oblique and made from a satellite to the east of the observatory.

Original satellite reconnaissance photograph by KH-9 camera (Mission 1215-3; June 19, 1979) available from the U.S. Geological Survey; photograph identification, interpretation, and processing by Mike Gruntman.

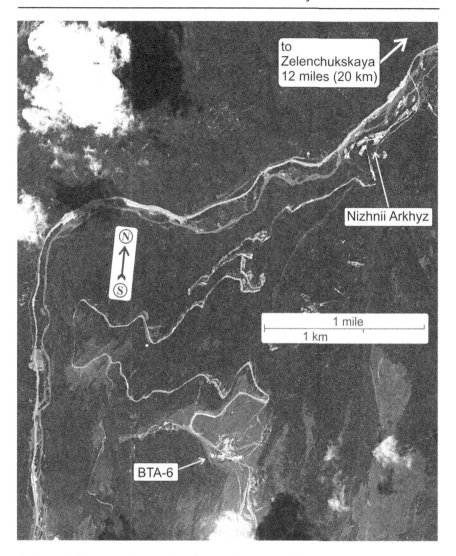

Fig. 5.12. Satellite photograph of the twisty mountain road connecting BTA-6 with the residential housing of the observatory staff at the supporting settlement Nizhnii Arkhyz (43°40.6' N, 41°27.4' E), 800 m (2600 ft) lower. A left tributary of the Kuban River, Bolshoi Zelenchuk, flows by the settlement on its course from the origins in the mountains to the south on its way north toward the town of Zelenchukskaya, 12 miles (20 km) away. The automobile road runs along the river, as is common in mountainous regions. Original satellite reconnaissance photograph by KH-9 camera (Mission 1215-3; June 19, 1979) available from the U.S. Geological Survey; photograph identification, interpretation, and processing by Mike Gruntman.

5. PSD: First Image and Beyond　　　　　　　　Mike Gruntman

Fig. 5.13. Oblique satellite photograph of the radiotelescope RATAN-600 (43°49.5' N, 41°35.2' E) with its reflecting mirrors forming a circle 600 meters in diameter. As noted in the preface of the book, distortions in oblique images have not been corrected. A horizontal line in the lower part of the circle is a plane reflector. The 12 radial lines are rail tracks for moving the cabins with signal receiving equipment. Original satellite reconnaissance photograph by KH-9 camera (Mission 1215-3; June 19, 1979) available from the U.S. Geological Survey; photograph identification, interpretation, and processing by Mike Gruntman.

Fig. 5.14. Reflectors of the Radioastronomical Telescope RATAN-600 near Zelenchukskaya at the Greater North Caucasus in October 1988. Photographs courtesy of Mike Gruntman.

research institute, *Vsesoyuznyi Nauchno-Issledovatel'skii Institut Televideniya* (VNIIT), or *All-Union Research Institute of Television*, in Leningrad, joined the cooperation. VNIIT brought expertise and the capability of building imaging sensors for defense and space applications—as well as more slowly moving, cumbersome bureaucracy.

Visits to SAO and interactions with astronomers were always interesting and educational. Suddenly, my expertise in nanosecond time-of-flight techniques for ENA detection pointed to potential new applications. We realized that some astrophysical processes, such as the falling of an interstellar dust particle onto a black hole, could produce possibly detectable short bursts of photons. Today, studies of rapidly changing intensities have evolved into a field known as time-domain astronomy.

Victor Afanasiev also brought into collaboration Romas Drazdis from the *Physics Institute* of the Academy of Sciences of Soviet Lithuania in Vilnius. Romas's group specialized in astronomical electronographic cameras, and their technologies and expertise were helpful for our development. (In electronography, each electron produced from a photocathode of an imaging system is accelerated to high energy and excites and forms a developable grain in a photo emulsion.) So, the cooperation grew.

Astronomers in SAO and engineers in VNIIT were excellent specialists. But despite all energy, excitement, and support, the work progressed glacially slow, as large organizations were accustomed to.

The foray into astronomy led to another interesting program. A group in the *Byurakan Astronomical Observatory* of the Armenian Academy of Sciences approached us in 1986 for help in the development of a focal plane photon-counting imaging detector for a planned improved version of their space-based ultraviolet orbiting telescope. The Byurakan observatory deployed a telescope, *Glazar*, for an ultraviolet survey of the sky in the 152-176 nm wavelength range at the astrophysics module *Kvant* on the manned space station *Mir*.[21] The telescope had a respectable 40-cm (15.7 in.) diameter mirror and an MCP-based image intensifier in the focal plane with its output photographed on a conventional film. Cosmonauts brought the film back to the ground for processing.

The Byurakan observatory planned a new similar-sized telescope on a dedicated Earth-orbiting astronomical satellite. Consequently, I did a feasibility study of replacing their image capturing system with an MCP-based photon-counting PSD in the focal plane. We rushed putting the results of this work together and published it[22] only weeks before I sneaked through the cracks in the Iron Curtain and disappeared forever. With the disintegration of the Soviet Union, the ambitious space-telescope program faded away.

To the best of my knowledge, nobody else had made MCP-based position-sensitive detectors in the USSR or other Soviet-block countries before I left in March 1990. My former colleague Kalinin and his external collaborators seemed to build a PSD for collisional laboratory experiments in the late 1990s. No publications in the refereed physics journals followed.

No Soviet or Russian position-sensitive detectors on the basis of microchannel plates were sent to space in the 1980s and early 1990s. During that time perhaps three dozen such PSDs flew on U.S. and European spacecraft and sounding rockets. IKI could have been building and operating such detectors in space instruments since 1983 if the leadership of the institute and my Department No. 18 supported it and allowed me to grow.

The political and economic system, however, has prevailed. These were the constraints, part of the broader "boundary conditions" that controlled and shaped our life and work.

---

21. Tovmasyan et al., 1988
22. Gruntman and Mkrtchyan, 1990

6. Boundary Conditions of Life and Work

Visitors in the land of real socialism

The boundary conditions imposed by the totalitarian state shaped the professional and personal lives of its subjects. The system inevitably suppressed the personal freedoms of everyone, denied dignity to human beings, and led to stagnation. Throughout history, however, people have adapted their lives to all environments. The Soviet Union was no exception.

The huge administrative state functioned wastefully and inefficiently but with some twists that sometimes provided openings. The founders (Fig. 6.1) of Marxist-Leninist hell on Earth, or paradise in view of fellow travelers living outside in the free world, designed the system that prevented, or at least minimized and controlled, interactions with foreigners. The bureaucratic behemoth of the land of real, mature socialism had an aberration, however.

The USSR Academy of Sciences ran a well-established and smoothly operating process of inviting foreign scientists to visit the country as its guests. Typically, a visitor covered his or her airfare, and then the Academy of Sciences picked up all expenses inside the Soviet Union. It took several months to get all the authorizations for such a visit, starting with approvals in the home department and by directors of the institute and culminating with the decision of the Academy's Foreign Relations Department.

The latter bureaucratic body most likely served as a front for the KGB and military intelligence engaged in scientific espionage in addition to performing ideologically sensitive administrative functions of control of interactions with foreigners. These tasks had been exercised for generations. A former Soviet scientist observed decades ago that the foreign section of the Academy of Sciences was staffed by "a strange mixture of people—some

6. Boundary Conditions   Mike Gruntman

Fig. 6.1. Soviet Union founder Vladimir Lenin (left) and his successor Joseph (Iosif) Stalin (right) ruled the vast country from 1917–1953. Removed from public streets and squares after the collapse of the inhumane socialist regime in the USSR, the statues stand today in an arts park museum in Moscow. Photographs (2018) courtesy of Mike Gruntman.

from the Foreign Affairs Ministry, some from the Ministry of Internal Affairs ... and some who had mistakenly chosen a scientific career, only to realize that they were totally unprepared and unfit for it."[1]

I was usually very open with visiting foreign colleagues about the peculiarities and realities of life in the country. Interestingly, some, but not many, West German colleagues understood perfectly well the Soviet system because of the recent history of their country. Many elements of administrative bodies, institutions, and practices of ideological siblings, National-Socialist Germany and the communist Soviet Union, exhibited similarities—two sides of the same radical socialist coin.

Foreign guests had diverse ideological views. Many, but certainly not all, visiting Western European and American scientists self-described themselves as "left-of-center" politically. A few adhered to solidly illiberal socialist political orientations. The latter fraction has been growing under

---

1. Klochko, 1964, p. 8

the influence of systemic leftism in the educational system in the West.

A British novelist and scientist, Charles P. Snow, observed in his famous lecture at the University of Cambridge in 1959 that "pure scientists still, though less than twenty years ago, have statistically a higher proportion in politics left of center than any other profession: not so engineers, who are conservative almost to a man."[2] Already by the early 2000s, common sense and rationality had departed some of my old friends and colleagues, particularly in Europe. They then experienced acute Bush[3] Derangement Syndrome. This irrational condition and accompanying intolerance markedly worsened when President Donald J. Trump took office in 2017. As for engineers, they have also been shifting to the left since Snow's assessment of the late 1950s,

Today, statistics reveal that "[p]arty voter registration of faculty in many professional schools [in American universities] is tilted overwhelmingly left ... in a country with the electorate evenly divided between the two main political parties."[4] On average, the Democrat-to-Republican registration ratio usually exceeds a factor of ten in elite colleges and universities. Hard sciences such as physics, mathematics, and chemistry fair a little better with "only" 5-6 Democrats for every Republican on the faculty. Engineering still stands out with the estimated 1.6 registration ratio.[5]

The non-hard-sciences and non-engineering fields such as "ideological" liberal arts and humanities are especially extreme in this imbalance, which inevitably degrades the quality of scholarship and replaces education with destructive indoctrination. The environment is becoming more intolerant and illiberal. A recent report by a U.S. President's Commission observed that the colleges in the United States had been peddling "resentment and contempt for American principles and history alike" and advanced "deliberately destructive scholarship."[6]

At the same time, a certain fraction of scientists and engineers always embraced and continue to support individual freedoms, a market economy, representative democracy, and human rights. Political and military leaders, the statesmen in the free world, valued and appreciated their contributions.

In the 1950s and 1960s, outspoken U.S. Air Force General Bernard Schriever oversaw the development of the first intercontinental ballistic missiles and played an important role in the early national security space

---

2. Snow, 2013, p. 33
3. U.S. President George W. Bush
4. Gruntman, 2018, p. 9
5. e.g., Langbert et al., 2016; Langbert, 2018
6. 1776 Report, 2021, p. 18

programs.[7] He noted in 1959 that "the disciplined and resourceful mind which makes possible intellectual performance of the highest order" was needed "for national survival" in the Cold War. This observation remains true today. Schriever also praised the contribution of intellectuals:

> In my view it is a national disgrace that the term "egghead" as a synonym for intellectual excellence has become a derogatory expression. Let me tell you that it is the "eggheads" who are saving us—just as it was the "eggheads" who wrote the Constitution of the United States. It is the "eggheads" in the realm of science and technology, in industry, in statecraft, as well as in other fields, who form the first line of freedom's defense against communism.[8]

Meeting visiting foreign colleagues provided an excellent vantage point to observe the leftward transformation. Already in the distant 1980s, the Western mainstream media was turning increasingly partisan left. Consequently, the Soviet Union did not look evil in the eyes of many of our guests, and some viewed socialism with sympathy and admiration.[9] My factual descriptions of the realities of life resonated with some and did not fit into the ideological preconceptions of others. A few visitors also expressed particularly hostile and hateful, bordering on antisemitic, views of Israel commonly propagated by the radical political left worldwide.

Some guests liked to concentrate on the shortcomings of their own countries. Criticism of the free world and self-loathing remains prominent today among intellectuals in democracies, especially on university campuses and in the media. Lack of perspective became a defining trademark of "progressives" and socialists. It originates in the failing educational system and leads to increasing ignorance and intolerance.

Similar to the prerevolutionary Russian intelligentsia, the intellectuals in many countries helped undermine and bring down their non-perfect political systems. They strived to replace them with a more just, as they believed, order. And similar to the Russian historical experience, the new regimes physically liquidated them and their families later, after they had served their purpose. Only a lucky minority succeeded to flee abroad where some would continue to advance the worldview that had brought the demise of their countries and their personal destruction in the first place. Such developments repeated themselves throughout history again and again in China, Cuba, Iran, and other places with devastating consequences and horrific human costs.

---

7. e.g., Gruntman, 2004, pp. 229-232, 236.
8. Schriever, 1959, p. 37
9. e.g., Kengor, 2010

Prominent dissident and human rights activist, the founding chairman of the Moscow Helsinki Group who spent nine years in harsh prisons and exile, physicist Yuri Orlov summarized the consequences of advancing socialism:

> Today, the brilliant and well-articulated idea of a just and rational society—the idea of socialism—has sustained a humiliating defeat. ... "Just" and "rational" Soviet society buried itself in an ocean of blood, and then took three decades to swim [back] to the surface again. Sixty-five million dead.[10] *Sixty-five* million. Any future society based on an extremist conception of justice and rationality [of socialism as advanced by intellectuals], on a final solution to social problems, will come to the same end [of the destruction of human beings].[11]

The so-called "democratic socialism" serves as a sinister gateway ideology to its inhumane hard varieties.

Another belated discovery awaits today's university professors enamored with hard-left orthodoxies: consistently implemented socialism does not tolerate academic freedom or recognize the privileges of the tenure. Many colleagues in academia have failed to notice this obvious inherent feature in the historical experiences of Russia, Germany, and China in the 20th century. Have they heard about the principle of karma? This is a rhetorical question.

Many years after reaching the United States, I attended a lecture by one American physicist who would be awarded a Nobel prize. He talked about his cooperation with a physics group in Moscow during the Soviet times. On one of his visits, he met a junior Soviet scientist who did not share the left-leaning worldview of the American guest. The condescending description of this young Moscow physicist who faced existential threats at that time, in stark contrast to the comfortable life of the visitor, struck me as particularly disgraceful. It displayed a combination of his political ignorance, arrogance, and lack of elementary empathy for the human condition.

Some like to explain such attitudes away as well-meaning naivete and gullibility. Facts do not support this generous assessment. A faculty member at a leading institution of higher learning in the free world cannot claim ignorance as an excuse.

On a lighter note, I often demonstrated to our foreign guests that the cumbersome all-controlling state did not enforce some rules. While strolling in the Moscow downtown, we could cross a street on the red light next to the KGB headquarters building (Fig. 6.2) in full view of several uniformed

---

10. Orlov refers here to one of the estimates of the number of lives claimed by Marxist atrocities.
11. Orlov, 1991, p. 8

6. Boundary Conditions — Mike Gruntman

Fig. 6.2. Symbol of oppression and repression by the Marxist state, the headquarters of the Committee for State Security in downtown Moscow in March 2018 (right). The statue of KGB founder Felix Dzerzhinsky (left) has stood in the middle of the square in front of this building since December 1958. In October 1991, the city of Moscow moved the statue to an arts park museum where it now shares space with other "retired" sculptures of the Soviet era. Photographs courtesy of Mike Gruntman.

police officers nearby. Nobody ever reprimanded or stopped us.

The realities of life in the Soviet Union also occasionally offered educational moments to the guests. A trip to Leningrad and then Tbilisi, Georgia, by Mike Lampton and his wife Susan provided such an example in 1983. First, customers in a coffee shop near the Hermitage Museum in Leningrad showed suspicion toward "capitalist" foreigners, got angry, and threw at us harsh words in foul language when Susan took a photograph of a colorful salad displayed on a stand. Then, an alert police officer "thwarted" an attempt to photograph a "strategic installation" at a passenger hall at the Pulkovo airport, where we waited for a late-night flight to Tbilisi. Mike took out his camera to photograph an iconic Russian samovar in a cafeteria.

We got away from this "offense" after I had a long conversation with the police shift supervisor. As I gathered from our friendly exchange, he was not happy to deal with this unexpected "international incident" interrupting activities planned for the night. Fortunately, he looked forward to watching a soccer game on a television set in his office followed by a romantic visit of his girlfriend. After promising him that taking photos would never ever

Fig. 6.3. Entrance to the main building of the Georgian Academy of Sciences in Tbilisi in 2014 (left). The coat of arms (right) of the Academy is above the door. Photographs courtesy of Mike Gruntman.

happen again, he released us to the dismay of his alert patrolman. The latter likely imagined a promotion for thwarting a hostile foreign action.

Then in the hospitable Tbilisi, the authorities canceled at the last moment a planned appearance of Lampton on the Georgian television. The reporters wanted to hear about Mike's preparations for a flight on the Space Shuttle. Global politics intervened here. A few weeks before the visit, President Ronald Reagan had challenged the United States to develop a missile defense shield in his famous speech in March 1983.[12]

Following a quick denunciation by supreme Soviet leader Yuri V. Andropov, the USSR Academy of Sciences harshly criticized the Strategic Defense Initiative in a statement[13] published in the most important official newspaper, *Pravda*. Numerous academicians obediently signed this "appeal to all scientists of the world" to undermine the Reagan initiative. Characteristically, many among them had been advocating and leading similar programs in the Soviet defense establishment for years.[14]

Soviet propaganda portrayed the Space Shuttle as a weapon system. Hence, the authorities canceled a television appearance by Lampton, a future Shuttle astronaut. At the same time, Mike's science lecture drew an attentive audience at the packed conference hall in the main building of the Georgian Academy of Sciences (Fig. 6.3).

12. Reagan, 1983
13. Appeal to all..., 1983
14. Gruntman, 2015a, pp. 5-8, 248-250

My boss Leonas usually encouraged and supported suggestions for inviting colleagues from abroad to establish scientific contacts. I had to "volunteer" to do the paperwork and logistical arrangements. Many IKI colleagues enjoyed opportunities to meet and interact with Western scientists.

As part of a visit by a foreign guest of the Academy of Sciences, it was possible to include a hop for a couple of days to other institutions, including, if appropriate, such interesting (for physicists) places as SAO in the North Caucasus or the Byurakan Observatory in Armenia. Having collaborators at these destinations always helped. The visitors typically gave seminar lectures on the latest developments in their fields, and local hosts benefitted from establishing new science contacts. The trips to observatories made visits interesting for our guests and sometimes included interactions in informal settings (Fig. 6.4).

A little bit more comfortable

Extensive collaboration required frequent travel to partners across the Soviet Union. Poor infrastructure and accommodations made trips unpleasant. However, visits to observatories in remote and picturesque locales stood out as exceptions. The astronomical centers had small hotels for visiting observers, which made trips unusually comfortable.

Getting a room in a hotel in any city in the country was exceptionally difficult for an ordinary person who did not belong to the nomenklatura caste. Some people were always more equal than others in a paradise of equity of real socialism.

Several times each year I had to go to Leningrad. These trips were among the worst, since getting a room in a hotel was next to impossible. Typically, one took an express train at midnight for an eight-hour journey in a sleeping compartment to Leningrad and spent the entire day there meeting partners and visiting various organizations. Then in the late evening one took another overnight express train back to Moscow and went to work straight from the station in the morning. The impossibility of taking a shower combined with legendary abusive poor services. But there was no choice for work to be done.

The Academy of Sciences paid for train trips in standard four-person sleeping compartments. If there were no tickets, then, "as an exception" and with approval by the cognizant deputy director of the institute, one could travel in a more comfortable two-person sleeping compartment. It cost twice as much.

Fig. 6.4. Participants of a neutral atom workshop (Space Gasdynamics Conference) visiting SAO in October 1988.

Top:
at RATAN-600, left to right, Stan Grzedzielski (Space Research Center, Warsaw), Gruntman (IKI), Peter Blum (University of Bonn), RATAN scientist host, and Hans Fahr (University of Bonn).

Left:
outing to local mountains; standing, left to right, Fahr, Grzedzielski, SAO scientist Yuri Balega, Blum, and the author of this book. Balega would succeed Afanasiev as director of SAO and serve in this capacity from 1993-2015. Since 2017, he has been vice president of the Russian Academy of Sciences. Photographs from collection of Mike Gruntman.

6. Boundary Conditions — Mike Gruntman

As noted in Chapter 1, the "as an exception" provisions permeated the system. Consequently, I always simply bought tickets for two-person compartments for my trips to Leningrad. On return, I wrote a memo stating the urgency of travel and unavailability of standard tickets and requested reimbursement of the more expensive fare. The story sounded plausible as "normal" non-nomenclature folks had difficulties in getting anything and everything.

My immediate boss Leonas obviously knew the real situation but never objected to my ruse. He made my life and work a little bit more comfortable by endorsing the reimbursement memos. Nobody wanted me to stop proactively going to our collaborators and switch to the glacial pace of "everybody else in socialism." The overseeing deputy director of IKI also always signed off on my requests. These were the state's funds, with little accountability, and he wanted the work to be done. Sometime in the mid-1980s, evening flights from Leningrad back to Moscow became convenient and practical, which also improved life.

Remote places in the wilderness

Vacations provided opportunities for recharging after exhaustive months. They lasted four weeks for scientists and engineers in the USSR. Vacations increased to six weeks for those with Ph.D. degrees and eight weeks for scientists with the next level post-doctoral D.Sc. degrees. Demand for recreational establishments, controlled and run as everything else by the government in socialism, always exceeded supply. At the same time, the general lack of accommodations and poor infrastructure limited individual travel, especially to popular resort areas.

Consequently, many scientists and engineers spent vacations in remote areas, trekking in the taiga, hiking in the mountains, or kayaking. Such time in the wilderness with several trusted friends provided a temporary escape from idiocy, pressure, and unrelenting propaganda of the Marxist-Leninist regime.

The mandatory Party-controlled rubber-stamp trade-union organization of IKI played some useful functions as well. In communist and other totalitarian countries, trade unions existed to control and manage the workforce rather than improve working conditions and raise salaries. Consequently, such organizations offered some educational, cultural, health, sports, and recreational programs instead of bargaining with the employers and defending the fundamental interests of employees. Many activities of Soviet trade unions resembled elements of Germany's *Strength Through Joy* (*Kraft durch*

Fig. 6.5. The author of this book (right) and the future IKI director Lev Zelenyi (left) at the institute's water sports camp on an island in a reservoir on the Volga River in the summer of 1979. Photograph from collection of Mike Gruntman.

Freude) organization of the 1930s. In IKI, the local cell provided kayaks to the institute's staff and supported a basic water sports camp in the summers. The enthusiasts maintained the latter on an island in a reservoir on the Volga River 100 km away from Moscow (Fig. 6.5).

My outdoor vacation adventures included numerous hiking trips in the mountains and taiga on foot, skis, kayaks, inflatable rafts, and even horseback (Figs. 6.6, 6.7, 6.8). The treks, combined, covered a few thousand miles in the wilderness and included nearly two hundred nights sleeping in tents.

Sometimes on such trips, we took with us officially looking but fake letters. They described our group as a "scientific-exploratory expedition" working on the development of a "multi-beam stellar interferometer" facility. I invented the fascinating name of the nonexisting out-of-this-world project. The letters requested all local authorities to assist us. Typed on the IKI letterhead, the text worked in some cases and helped get accommodations and bus tickets in remote areas of the country, with particularly poor logistics.

A few times, my vacation trips gave me an opportunity to surreptitiously explore the areas near state borders. I looked for ways to get out of the country.

A couple of ski trips in the Karelia and Murmansk regions in the European north of the country clearly showed the impossibility of getting to places in the vicinity of the border with Finland. Authorities restricted and tightly controlled access to such areas. Besides in the words of a knowledgeable British intelligence officer, this neighboring country, Finland, turned "over to the KGB any fugitives from the Soviet Union that fell

6. Boundary Conditions — Mike Gruntman

Fig. 6.6. Many IKI scientists and engineers enjoyed vacation travel in the wilderness away from the Marxist-Leninist "civilization." In a small group of close friends, one could spend relaxing time in remote areas as long as they were not in the vicinity of the state border.

The top photographs show the author of this book hiking (right) in rugged mountains in Tajikistan and diving (left) in Iskanderkul lake in the Gissar Range of the Pamir-Alay system in Central Asia. Trekking wilderness on skis in winters presented particular challenges because the commercially available touristic gear was primitive or nonexistent. The bottom photos show the author with a colleague on a winter trip in the Russian North, covering more than 100 miles on skis in the wilderness with heavy backpacks and sleeping in tents near the Arctic Circle in 1983. Such ski trips also helped improve survival skills in the northern wilderness in case an opportunity arose to cross the border. Photographs from collection of Mike Gruntman.

Fig. 6.7. The author of this book next to his kayak (left) and cutting wood for the campfire (right) during a spring trip in central Russia. Kayaking enjoyed special popularity during extended national holidays in early May. Many snow patches remained in the woods at that time of the year when the spring was late. Photographs from collection of Mike Gruntman.

Fig. 6.8. The author of this book with a saw (left) on a ski trip beyond the Arctic Circle preparing wood for the campfire, his favorite chore in the wilderness. Friends used to joke that these skills would promote him to a foreman of a prisoner tree-felling detail in a Gulag camp in Siberia when the authorities finally get him. Photograph from collection of Mike Gruntman.

into their hands... The Finns deeply resented the term 'Finlandization,' but it accurately represented the situation" during the Cold War.[15] Crossing additional hundreds of miles through Finland to the safety of Sweden or Norway presented a formidable problem.

Many years later, I read that Soviet physicist George Gamow had discovered a very similar environment in the same border region when he explored it for the same purpose in 1932. The local population was also allowed then, in his words, "to keep all the money paid to them for [help with] 'transportation' [across the border], plus an additional amount to be paid to them by the border officials into whose hands" they delivered those who attempted to escape.[16] In many regions near state borders, the authorities organized groups of schoolchildren, "young friends of border guards" and alike, to hunt strangers down.[17] The primary purpose of the huge border control forces was always stopping Soviet subjects from escaping their beloved country rather than guarding and protecting the land from smugglers and real or imaginary hostile capitalist saboteurs and spies.[18]

The idea of crossing the ice-covered Arctic Ocean on skies to Canada through the north pole looked truly fascinating. Unfortunately, this was unrealistic as an assessment of the required gear, maps, and firearms clearly indicated. On the positive side, I read all available semi-scientific and advanced popular books on polar bears and became rather knowledgeable about their populations and habits in the Arctic. I even found an error in one of the scholarly publications. I then mailed a letter to the author in the Academy of Sciences in remote hope of getting into his next science expedition to the far north in a supporting role. Without success.

One summer I went hiking alone on an exploratory trip in the mountains of Soviet Armenia. By luck, I met on a remote trail a plainclothes police detective working on narcotics trafficking. We shared food for lunch and had a leisurely friendly conversation for a couple of hours hiding in rock shadows during the hottest part of the day. He had a profound knowledge of the local geography and security regulations. A chat with a physicist from a distant academy of sciences working on space projects was a pleasant change of routine and company for him as well.

It became clear that there was no way of getting close to the state border with Turkey in Armenia. The local population also enthusiastically turned strangers in to the authorities. While I was very careful in my questions,

---

15. e.g., McIntyre, 2018, pp. 401, 402
16. Gamow, 1970, pp. 118, 119
17. Golovanov, 2001, v. 1, p. 157
18. e.g., Golovanov, 2001, v. 2, pp. 213, 334

my new "buddy" was a street-smart detective. He might have "smelt" my hidden agenda but did not show this. Who knows what was on his mind.

A couple of days later, hospitable local shepherds invited me to spend the night in their family summer camp. They had a few large yurts, hundreds of sheep, ferocious dogs, and delicious food. Among them, I looked like an alien with my tent and colorful sleeping bag.

My hosts belonged to a not-favored minority in Armenia, the Yazidis, who primarily populated areas near the Turkish border. The new friends were an excellent source of information on local conditions. They confirmed what I had learned from the police detective about the impossibility of approaching the state border by a stranger.

"Clandestine" advice

My expertise in microchannel plates and position-sensitive detectors led to unusual personal consequences. Heads of a few IKI laboratories and groups periodically asked me for help in understanding the designs and workings of foreign space instruments with MCPs. They either saw them in scientific journals or their foreign collaborators proposed MCP-based sensors for joint experiments on Soviet space science missions. More than once they asked me to look at their ideas for new instruments with such detectors and to help in evaluating testing of microchannel plates and channel electron multipliers.

Laboratory heads at IKI did not view the group of Leonas as competition because we did not have flight experiments. Therefore, talking to me was "safe" and I always respected the strict confidentiality of the interactions. The moment IKI included our ENA instrument on the Phobos mission (Chapter 9), the situation changed overnight for some of our peers. One otherwise friendly laboratory head made multiple trips to a deputy director of IKI calling to jettison our experiment from the payload. The deputy director then gleefully told about these requests to Leonas and Baranov.

A few times each year, Konstantin Gringauz asked me to stop by his office for "conversations." They took place always after 6 p.m. when most of his staff had left for home. He had doubts about some instrument development projects in his laboratory. I helped evaluate designs and work progress.

Rashid Sunyaev was interested in position-sensitive detectors, especially those built and flown by the Space Astrophysics Group of Stuart Bowyer in Berkeley. Knowing about my working relations with Mike Lampton from that group and Bowyer's good opinion about my work, Sunyaev invited me for private chats several times.

6. Boundary Conditions   Mike Gruntman

A scientist in another IKI unit consulted with me on technical issues but only when his laboratory head was away from the institute. The latter did not want anybody outside his group to know details of their projects and especially about encountered difficulties. Such discussions also led to the establishment of trusted relationships.

These "clandestine" activities opened direct access to heads of several leading scientific units at IKI, sometimes competing with each other. Such unusual (for a "lowly" junior scientist) access led later to offers to stay at IKI when Department No. 18 transferred to another institute in 1987 (Chapter 11).

Snafu in science diplomacy

In 1982, a space instrumentation workshop took place in a major regional center on the Black Sea shore, Odessa (Odesa), Ukraine. More than two dozen specialists from IKI and many scientists and engineers from other institutions across the country participated. An ocean-going ship, *Akademik Sergei Korolev* (*Academician Sergei Korolev*), which provided communications with civilian and military spacecraft outside the Soviet landmass, happened to be in the Odessa port at that time as well (Fig. 6.9).

Oleg Vaisberg (Fig. 6.10) and I shared a room in a hotel during the workshop. This arrangement likely occurred because laboratory head Vaisberg thought it convenient to have me as a roommate rather than a scientist from his group. Being from a different unit helped avoid awkwardness.

Parenthetically, sharing hotel rooms was common in the Soviet Union. Many hotels had rooms for several people or even more, housing total strangers in the same room. One was happy to find a bed in a hotel, a separate room being a true luxury reserved for the nomenklatura. Diners also shared tables in restaurants and cafeterias in a common-sense response to the shortage of basic necessities. Languages reflect the realities of life for their speakers, the people. Characteristically, the Russian language has not developed the word "privacy."

A group in the *Odessa Polytechnic Institute* organized the workshop. During the meeting, I established friendly relations with its sympathetic graduate students and young engineers. As I planned to leave the town one day later than most workshop participants, the organizers invited me to a party marking the conclusion of the workshop. The large scientific gathering went very well because of their hard work. Now it was the time to celebrate the mission being accomplished. At the party, they revealed a "big secret" of special attention to me.

Fig. 6.9. *Akademik Sergei Korolev* near Odessa in 1982. The ship supported space operations by providing tracking, telemetry, and control for civilian and military spacecraft from locations outside the Soviet landmass. Photographs courtesy of Mike Gruntman.

Fig. 6.10. Friendly dinner with Oleg Vaisberg (left) and his wife Yulia in Los Angeles in December 2015, 40 years after Oleg and I met for the first time at IKI. From collection of Mike Gruntman.

The contracts from IKI's laboratory of Gringauz supported much of the research and development work of the Odessa group. They built parts of space plasma instruments. The head of the group, a professor in Odessa Polytechnic, wanted to expand his collaboration to Gringauz's competitor, Vaisberg, who could also provide contracts. They thus needed to establish links with Vaisberg without offending their patron-saint Gringauz. And it would have been a "mortal" offense.

At the "war council," the Odessa group concluded that I must be among the most trusted and close to Vaisberg members of his laboratory, sharing the hotel room with him. Therefore, the energetic guys and cute girls from the Polytechnic Institute got marching orders to befriend me. The latter was not difficult, and we quickly developed true mutual sympathy. We roared with laughter and joked about our bosses when we figured out what happened.

For a few years, I had been buying razor blades and mailing them to my new Odessa friends. The country was building rockets, launching spacecraft to the Moon and planets, splitting the atom, and uncovering mysteries of the Universe. But it could produce blades to fill the shelves only in Moscow and Leningrad but not in other major cities.

It all could end in one moment

Sometime in 1984 or 1985, IKI deputy director Georgii P. Chernyshev summoned me to his office. Chernyshev, a KGB colonel, oversaw all security aspects of life in the institute. The security service placed its active-duty and reserve officers in high-level positions in all important and sensitive organizations in the country.

Before the assignment to IKI, Chernyshev "oversaw" the famed Bolshoi Theatre. As rumors had it, something went wrong in the Bolshoi, perhaps a defection, and the KGB sent him, in a reshuffling of the security personnel, to "strengthen" space science at IKI.

Sagdeev used Chernyshev's connections to the theater to obtain highly sought tickets to the Bolshoi ballet and opera performances for leading IKI scientists and administrators on the Day of Cosmonautics,[19] celebrated on April 12 each year. Ballistic missile and space establishments in the country usually recognized their specialists with awards and bonuses on that date. The Day of Cosmonautics commemorated the anniversary of the first human flight in space by Yuri Gagarin in 1961.

Chernyshev told me that "a couple of their [KGB] guys" would come the next day to the institute to talk to me. He did not give details. I had never had any personal interactions with Chernyshev before that day. He certainly knew much about me as it was largely his decision to bar me from foreign trips. No doubt that the informers occasionally mentioned my name in their reports in uncomplimentary ways.

Nothing concentrates attention more than a life-ending threat. I thought really hard about possible reasons for the KGB visit. My boss Leonas told me that he would be absent the next day, visiting a main national library. He offered his office for the meeting. It was a diplomatic way for him to stay out of potential troubles as well.

When I got home that night, I first hid the banned book *Animal Farm* by George Orwell. I worked until the wee hours searching through my bookshelves and burning some papers. It was not easy, as books packed the shelves in my room (Fig. 6.11). I practically did not sleep that night, trying to figure out where I could have slipped and who among my close friends might have turned me in.

The next day, two young KGB guys in civilian clothes, slightly older than me, came to the institute. They turned out to be scientists from an engineering unit working on sensors for revealing practically invisible fingerprints. Somebody pointed at me as the leading specialist in ultrasensi-

---
19. Gruntman, 2007a, Chapter 4

6. Boundary Conditions   Mike Gruntman

Fig. 6.11. The author of this book at home in front of bookshelves in the early 1980s. He urgently purged his book collection and documents when instructed by IKI deputy director to meet KGB officials. Photograph from collection of Mike Gruntman.

tive photon-counting imaging detectors. It was not surprising as my review article on PSDs had already appeared in *Pribory i Tekhnika Eksperimenta*. Instead of simply calling over the telephone as colleagues from other "normal" institutions routinely did, the KGB "guys" proceeded "through their special channels."

Approximately at that time, I heard on the Western radio that the Swedish domestic security service apprehended a local businessman illegally transhipping to the Soviet Union equipment that was banned for export by COCOM.[20] In addition, a large truck with European license plates recently unloaded boxes with a foreign computer in the IKI backyard. My friends in the institute told me, as a big secret, that it was an embargoed computer from Japan. It had had all origination labels removed from the external panels, but small manufacturer tags remained attached to subsystem boxes and many circuit boards inside.

I told the KGB visitors that they could buy commercially available position-sensitive detectors that had been advertised in Western physical journals such as *Review of Scientific Instruments* and *Physics Today*. I even suggested, emboldened and in jest, that their "KGB guys would undoubtedly find ways to snatch them" in case the detectors were forbidden for export.

---

20. *Coordinating Committee for Multilateral Export Controls*, COCOM, limited exports of sensitive technology to communist countries during the Cold War.

My visitors grinned and, without any hesitation, nodded in agreement that this would not be a problem at all. Even my use of a "not nice" word, "snatching," did not bring any objections. They considered me as one of them, I guess, a loyal team member.

This time, the event turned out to be a false alarm, just a scare. It reminded me, however, how fragile everything was. My life and work may literally end in one moment. Interestingly, I had hidden the book by Orwell so well that I could not find it later to give to a friend to read. Only after many years did I figure out where I put it.

The free spirit in the ether

The above-mentioned Western radio played an important role in the Cold War struggle between the free world and the dark forces of Communism. It kept free spirits alive in totalitarian Marxist hell.

The information brought by the radio mitigated, in part, fabrications in *Pravda*, other newspapers and magazines, radio, and television programs. Being continuously exposed to the output of the agitation and propaganda arms of the Communist Party led to the development of abilities to identify falsehoods. Therefore, it is so easy now to spot tendentious misrepresentations, a rampant bias of the "narratives," and outright lies in the dominating left-wing media in the United States. Europe is no different in this respect.

Today, the most influential American newspapers play the role of the Democratic Party press, the giant tech firms control the distribution of the news and manipulate and filter them, and the media, in general, turned hard left.[21] The term "mainstream [media]" has lost its meaning. My past experience makes it possible to read fake news stories in self-described newspapers of record and watch television networks and cable channels that are brimming with self-importance and still be able to figure out, despite distortions, what is really happening in the country and the world. Perhaps minute details are lost, but certainly the main story comes across.

Even in the closed society of the Soviet Union, there were ways of obtaining some banned information despite all restrictions, censorship, unending deceit, and brainwashing. Thinking rationally, testing hypotheses, and relying on experimental facts and observations, which is the basis of the practice of physics, supported the development of independent points of view. Common sense contributed mightily as well. The Western radio helped the cause of freedom in an important way, despite relentless attempts to block transmissions.

---

21. e.g., Groseclose, 2011; Silberman, 2021

Seventy years ago, the U.S. Ambassador to Moscow wrote that the Kremlin began "the extensive program of jamming" Russian-language broadcasts by the Western radio in the spring of 1949.[22] A contemporary CIA report pointed out that jamming of Russian-language programs from Madrid in Spain was conducted as early as 1946. Then, "[b]eginning on 24 April 1949, a new period of particularly intensive jamming of [*Voice of America*] VOA Russian-language frequencies began."[23]

Full-scale warfare in the ether had erupted.

The Marxist state could not tolerate independent sources of information and disapproved of the subjects who listened to foreign radio. Fortunately, the punishment usually was moderate and did not lead to imprisonment for this widespread practice. Already in the mid-1950s, a CIA report described that "listening to foreign broadcasts by the Soviet populace is generally done on the sly and while the individuals are alone. Home listening, within the strict family circle also seems to be a normal practice. Not knowing who can be trusted or who is sufficiently reliable, the families are careful not to create suspicion among their neighbors."[24]

Broadcasting and jamming

Several countries, a coalition of the willing, would broadcast to the communist world behind the Iron Curtain in many languages. The most prominent government stations included Voice of America, the U.K.'s *British Broadcasting Corporation* (BBC), West Germany's *Deutsche Welle* (German Wave), and Israel's *Kol Israel* (Voice of Israel).

In addition, two formally nongovernment sister stations, *Radio Free Europe* (RFE) and *Radio Liberty* (RL), were particularly active in highlighting the failures and crimes of the regimes. Originally, the Central Intelligence Agency funded RFE and RL in secret. Later, in 1971, the U.S. Government began open funding of the stations by Congressional appropriations. In turn, the Soviet Union and the People's Republic of China broadcast numerous communist radio programs to various countries across the globe.

In the mid-1980s, Voice of America provided "nearly 270 frequency hours daily into the Soviet Union and Eastern Block countries. Radio Free Europe and Radio Liberty broadcast more than 100 frequency hours."[25] In addition, Voice of Israel beamed more than 50 hours into the Soviet Union each day.

22. Smith, 1950, p. 182
23. CIA, 1954, p. 107
24. CIA, 1954, p. 107
25. Sowers et al., 1988

Initially, in the 1940s and 1950s, some broadcasting was in the medium frequencies but then shifted almost exclusively to high frequencies or short waves. The most common wavelengths were at the 13 m range (frequency band at 23 MHz), 19 m (19 MHz), 25 m (12 MHz), 31 m (9.5 MHz), 41 m (7 MHz), and 49 m (6 MHz). Broadcasts at the shortest wavelengths, 13 m, 16 m, and 19 m, became active only in the 1970s and 1980s as it was difficult earlier to find radio sets in the Soviet block capable of reception in those ranges.

The Soviet state invested huge resources in jamming foreign radios, which with time became a major and expensive undertaking. It combined local transmitters that emitted ground-wave (or direct-wave) countersignals in the area of jamming and distant transmitters relying on sky-wave propagation of their emissions. The latter signal propagation relied on reflection of radio waves from the ionosphere, a layer of the ionized upper atmosphere at altitudes of a few hundred kilometers above the ground.

The conditions in the ionosphere and the corresponding quality of radio transmissions depended on the time of the year, time of the day, geographical latitude, and solar activity. Specialists noted that "[l]ocal jamming is very cost-effective in areas that are large population centers such as major cities. Sky-wave jamming is a cost-effective method of disrupting broadcast services over large areas."[26]

The Soviet government built numerous powerful transmitters in the 6-25 MHz frequency range. They broadcast Marxist propaganda programs to foreign countries across the globe relying on sky-wave signals reflecting from the ionosphere. The same radio centers sent jamming signals to cover the interior of the Soviet Union.

The propagation distance for radio signals with a single reflection from the ionosphere varied from 500-3500 km (300-2200 miles). Therefore, sky-wave jamming called for important international cooperation to cover the desired areas beyond national borders and within neighboring fraternal socialist countries. A total of two dozen sites with powerful transmitters across the Soviet Union and Eastern Europe waged war in the ether. Most sky-wave jamming emissions were accompanied by two-character markers in Morse code. This allowed jamming efficiency to be monitored.[27] Numerous local direct-wave transmitters complemented sky-wave jamming in individual cities.

The Soviet Union has been developing the science and engineering of sky-wave propagation for many years. This technology had another

---
26. Sowers et al., 1988
27. Sowers et al., 1988; Pleikis, 2002

important application and formed the basis of the over-the-horizon radars. Such radars, named *Duga*, operated at frequencies 5-28 MHz (wavelength 11-60 m) and played a major national-security role, supporting early detection of intercontinental ballistic missile launches.[28]

Two Politburo members, Secretary of the Central Committee of the CPSU Yegor K. Ligachev and chairman of the KGB (in 1982-1988) Victor M. Chebrikov, summarized the state of jamming in the Soviet Union in 1986 in a memo to the Central Committee of the Communist Party:

> Thirteen radio centers for long-distance [sky-wave] defense [jamming] and 81 stations for near defense [local jamming] are being used with a total power [of transmitters] of about 40 thousand kW [40 MW].
>
> Long-distance defense provides jamming of broadcasts on approximately 30% of the territory of the Soviet Union. Stations of local defense [jamming] are deployed in 81 towns and suppress broadcasts in areas with a radius up to 30 km [20 miles] [around the jamming stations]. Beyond that zone, the quality of jamming sharply decreases. Means of long-distance and near defense cover with varying degrees of efficiency the regions of the country with about 100-130 million people [out of the population of 250 million].[29]

Occasionally, the Soviet Union suspended jamming of Russian-language broadcasts of Voice of America, BBC, and German Wave, notably from 1963-1968 and 1973-1980. Such steps usually supported Soviet foreign policy objectives and reflected international developments. For example, the latter quiet period ended in 1980 with the rise of the Solidarity trade union movement in Poland that threatened communist domination. Whenever jamming of selected foreign radios ceased, the transmitters switched to additional suppression of Radio Liberty and Voice of Israel.[30]

Some knowledge of foreign languages helped overcome jamming as well. Soviet authorities interfered with Russian-language programs especially effectively in the capital of the country. At the same time, one could often receive in Moscow the Voice of America and BBC broadcasts in English with decent quality.

The political and cultural left had not overtaken these influential American and British radio stations in the 1970s and 1980s. They then provided factual information, offered a balanced view, and advocated causes of freedom, representative democracy, a market economy, and human rights.

---

28. Gruntman, 2015a, pp. 242-244
29. Ligachev and Chebrikov, 1986
30. Pleikis, 2002

In short, at that time the radios fought in the Cold War on behalf of the free world.

Forbidden fruit of foreign broadcasting

The Communist Party considered jamming Radio Liberty the highest priority. Therefore, the broadcasts of RFE/RL were never afforded a reprieve and "had been the target of more than 70 percent of all Soviet jamming."[31] At times, one could not practically receive Radio Liberty in Russian in Moscow at all. However, one could often clearly listen to its sister service, Radio Free Europe, in Polish unmolested. The latter obviously did not have many listeners in the Soviet capital.

Director of Central Intelligence Allen Dulles enthusiastically supported an expansion of Radio Liberty broadcasting in the 1950s. Dulles noted that a CIA study demonstrated that "the Soviets had made enormous strides toward eliminating illiteracy and improving the technical education available to their citizens." He shrewdly observed then that

> however narrow and ideologically controlled was the education being offered, it was bound to stimulate the imagination and curiosity of millions of Soviet citizens... [A]nd they would inevitably begin to ask questions and then to listen to foreign broadcasts when dusty answers of the regime left them unsatisfied. If we [CIA] could override the jamming with consistently accurate reports as what was really going on inside their country and in the outside world, we could look forward to the day when a mass audience would dare to eat the forbidden fruit of foreign broadcasting.[32]

Initially, RFE and RL broadcast from West Germany. To counter Soviet jamming, Dulles authorized the construction of powerful sky-wave radio transmitters for Radio Free Europe in Portugal and for Radio Liberty 500 miles (800 km) to the west at Playa de Pals on the Mediterranean shore 85 miles (135 km) from Barcelona in Catalonia, Spain (Fig. 6.12). They became operational in 1952 and 1959, respectively.

A government report to the U.S. Senate described that

> Radio Liberty's first broadcast to U.S.S.R. on March 1, 1953, from a single station at Lampertheim [near Mannheim in Hesse, West Germany], was with one shortwave transmitter of 10 kilowatts and in one language, Russian. Since then Radio Liberty reportedly has become the most powerful free voice heard in U.S.S.R.

---

31. Sosin, 1999, p. 201
32. Meyer, 1980, p. 116

6. Boundary Conditions — Mike Gruntman

Fig. 6.12. Transmitters of Radio Liberty on the shore of the Mediterranean Sea on the Costa Brava north of Barcelona in Catalonia, Spain, in the 1960s. The site housed six transmitters with 250 kW power each. Four transmitters could also be linked together to send a megawatt signal. Reflected from the ionosphere, skywave signals effectively reached densely populated parts of the Soviet Union, with Moscow being about 3,200 km (2,000 miles) away. In the late 1970s, "[d]espite the severe jamming ... Radio Liberty listenership averaged 7.6 million people a week" (Sosin, 1999, p. 173). These antennas of Radio Liberty were demolished on March 22, 2006. Photograph credit: RFE/RL, USAGM.

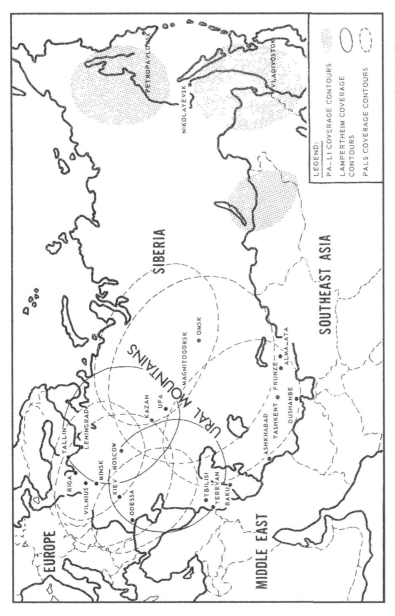

Fig. 6.13. Coverage of the Soviet Union by sky-wave signals of Radio Liberty's transmitters at Lampertheim in West Germany (solid contours), Playa de Pals in Spain (dashed contours), and on Taiwan (dot-filled regions). Map: Comptroller General..., 1972, p. 40.

6. Boundary Conditions Mike Gruntman

In July 1971 Radio Liberty had 17 transmitters at locations in FRG, Spain, and Taiwan, broadcasting 295 transmitter hours a day in Russian and up to 18 other languages of U.S.S.R. Radio Liberty's combined power totaled 1,830,000 watts.[33]

Radio Liberty leased "the transmitter station on Taiwan from the Broadcasting Corporation of China." After improvements and modernization, the site took advantage of "overwater reflection for signals directed to the U.S.S.R. Far East. Radio Liberty operates three 50,000-watt transmitters."[34] It beamed RL programs from Taiwan since 1955.

The Lampertheim site in Germany had also increased its capabilities to six 50-kW transmitters by the early 1970s. It could also transmit programs from RL's Munich studios to Playa de Pals if needed. The latter station in Spain broadcast programs by six transmitters with 250 kW power each (Fig. 6.12).

Figure 6.13 shows the footprint of coverage of the Soviet Union by sky-wave signals of Radio Liberty's transmitters at Lampertheim in West Germany, Playa de Pals in Spain, and on Taiwan. The broadcasts reached major population centers of the country. RL and its partner RFE effectively carried the truth about the world to the closed societies of communist countries for decades.

Today, in the words of Michael Pack, the former chief executive officer of the overseeing *U.S. Agency for the Global Media*, USAGM, foreign broadcasting increasingly follows the lead of the mainstream media that have transformed "away from the old standards and objectivity and balance" and became "nakedly and unapologetically partisan and networks like VOA follow their lead."[35] It is a true shame.

The left turn of the U.S. media is converting its dominating parts into an agitprop arm of the Democratic Party marching toward Marxism. As a result, the media has lost much of its credibility. A commentator in the largest national newspaper observed that "[i]f the American media says in unison that something is the right thing to do, nearly half the country [United States] takes that as proof it is the wrong thing to do."[36] This development also inflicts irreparable damage to human rights worldwide and directly harms those brave souls standing for freedom and against oppression in various parts of the globe.

Freedom is not free.

---

33. Comptroller General..., 1972, p. 39
34. Comptroller General..., 1972, p. 121
35. Pack, 2021
36. Henninger, 2021

CPSU and fellow travelers fight back

The Communist Party of the Soviet Union did not remain idle in the existential ideological struggle. Every night, the main and only Soviet national television channel beamed the evening newscast, *Vremya* (Time). The program began with appearances and speeches of the Politburo members followed by reports on decisive successes in battles for harvesting crops and producing ever more steel and coal. Main international news stories were next. Then each day, a few fellow travelers from foreign countries paraded: Western intellectuals, professors, and writers; trade union leaders; and assorted Marxist-leaning rulers and activists from the Third World.

Some of these supporters visited the Soviet Union at that time while others opined from the comfort of their home countries. They all enthusiastically praised the socialist system in the USSR and highlighted the despicable "ulcers of capitalism"[37] permeating the free world. Notably, leaders of the American anti-communist AFL-CIO trade unions practically never appeared in the broadcasts.

The Soviet Union has been nurturing anti-capitalist and anti-American elements in the West for decades. Communist intelligence services planted numerous false stories in the receptive left-wing media, using disinformation as an effective propaganda tool against the free world.[38] These operations were part of "active measures" in their trade parlance.

A senior Soviet intelligence officer described that,

> [u]nder [KGB Chairman, 1967-1982] Andropov the disinformation branch of KGB flourished. For both domestic and external consumption, it concocted stories to deceive, confuse, and influence targeted audiences. It conducted operations to weaken Soviet adversaries and to undermine internal stability and foreign policies of the Western world to facilitate favorable conditions for the eventual triumph of Communism.[39]

That some leftist intellectuals were gullible dupes[40] did not diminish their responsibility for the horrendous human cost of their complicity. An intellectual cannot claim ignorance as an excuse.

There were also professional pro-Soviet stooges. Anatoly Chernyaev, a life-long high-level CPSU official and senior advisor (from 1986) to Mikhail Gorbachev, wrote with disdain about the Soviet-controlled "peace

---

37. "Ulcers of capitalism" was one of a dozen stock phrases of communist propaganda widely used by the media and in the educational system in the USSR.
38. e.g., Bittman, 1985; Pacepa, 1987; Pacepa and Rychlak, 2013
39. Kalugin, 2009, p. 297
40. e.g., Red visitors..., 1949; Kengor, 2010

movement" in an entry in his private diary. Senior representatives of the communist parties of Bulgaria, Czechoslovakia, Poland, East Germany, and USSR discussed

> financing the [World] Peace Council [WPC], WFTU [World Federation of Trade Unions], and others, as they have budget deficits each year, [with expenses] exceeding the allocated budgets each year. They have been spending on mistresses, various "events," travel, and luxurious life—professional fighters for peace.[41]

The cheerful participation of the international left in the propaganda war against Soviet people and their brainwashing directly contributed, wittingly or not, to oppression and massive repressions in the communist world throughout decades.

As early as 1939, a former top Soviet intelligence officer, Walter Krivitsky, brought this fact to the attention of U.S. Congress. Krivitsky had defected in 1937 rather than, in the words of Allen Dulles, "return to Russia to be swallowed up in the purge." Then in 1941, he "was found dead in a Washington hotel ... shot presumably by agents of the Soviets who were never apprehended."[42]

Krivitsky testified to Congress that

> if it were not for the work of the Communist Party [of the Soviet Union] abroad, and of the sympathizers [in the West] during 1936 and 1937, the purge [in the Soviet Union] could have never been carried out successively. This demonstrates very clearly why Stalin needs them... It is evident that he [Stalin] required some support from abroad for all these measures [purges].[43]

Krivitsky also emphasized that "[f]oreigners little realize how vital it was" for Stalin and CPSU to declare unanimous support of the physical liquidation of the opposition and how important hailing the Soviet constitution was "by many foreign liberals as, if not a great achievement, at least a 'significant aspiration.'" He also specifically singled out, as an example, that the help that renowned French author and Nobel Prize winner in literature Romain Rolland "gave to totalitarianism by covering the horrors of Stalin's dictatorship with the mantle of his great prestige, is incalculable."[44]

The complicity of the radical left and intellectuals in the West did not begin with the support of the purges of the late 1930s in the Soviet Union.

---

41. Chernyaev, 2008, entry December 14, 1974
42. Dulles, 2006, p. 135
43. Krivitsky, 2004, pp. 66-67
44. Krivitsky, 1939, pp. xiii, 61, 69, 70

It went back to the earlier days of the Marxist paradise.[45] Covering up the communist-caused great famine of 1932-1933 in Ukraine (known today as the *Holodomor*), Kazakhstan (still disregarded by many), and other parts of the country is a prime example of the consequences of the ideologically-driven bias of the Western media. Honest reporting could have saved millions of lives. The same actors provided later support and comfort to the actions of the Communist Party of China with an even greater human cost.

A prominent U.S. diplomat, George Kennan, observed that the Kremlin realized early that for "the groups of bourgeois-liberal enthusiasts ... the Soviet Union soon came to have so powerful an attraction. Soviet policy thus began with time ... to concentrate on using all foreign sympathizers, Communist or otherwise, as vehicles for a purely nationalistic Soviet foreign policy."[46]

Suppression of dissent and repressions in totalitarian countries such as the Soviet Union, the People's Republic of China, and Cuba always relied and continue to depend to this day on the help of sympathizers and front organizations abroad. The U.S. Senate succinctly summarized that

> Communism has claimed the lives of more than 100,000,000 people in less than 100 years.[47]

In the 20th century, the Marxist regimes lasted longer because of the passionate support by a significant part of the international left. In addition, many liberals rejected muscular anti-Communism and remained equivocal about bloody repressions. Millions of lives would have been spared if it were not for the fellow travelers.

The brutal carnage was not the only "achievement" of the international left that also contributed to the calamity of World War II. A keen British observer noted that "the intellectual sabotage [of the national morale and preparedness in Great Britain] from the Left was partly responsible" that "the Fascist nations judged that ... it was safe to plunge into [world] war."[48] Not surprisingly, supposedly intelligent scientists praised and celebrated the Munich agreement.[49]

At the same time, as noted by a high-ranking U.S. diplomat Robert Murphy, in another country, France, Marxists followed the Moscow lead and mightily contributed to breaking "down the morale of the French fighting forces," leading to the military defeat in 1940 and the tragedy of occupation.

---

45. e.g., Courtois et al, 1999; Kengor, 2010
46. Kennan, 1983, p. 517
47. Senate Resolution, 2005
48. Orwell, 2005 (1940), pp. 40-41
49. Promotion of peace, 1938

6. Boundary Conditions   Mike Gruntman

Fig. 6.14. Top: monument, *Victims of the Totalitarian Regime,* in an arts park in Moscow. Bottom: *Memorial for the Victims of Political Repression during the 1930–1940s and Early 1950s* at the location of the Gulag ALZhIR prison camp near Nur-Sultan (former Astana), the capital of Kazakhstan. The acronym ALZhIR stands for the *Akmolinsk Camp of Wives of Traitors to the Motherland.* Akmolinsk is the old Soviet name of the city Nur-Sultan (Astana). Photographs (2018 and 2016, respectively) courtesy of Mike Gruntman.

Fig. 6.15. Large sculptures, left-to-right, of "banished" Joseph Stalin, Vladimir Lenin, and bigger Stalin with the outstretched right hand in Tirana, Albania. Photograph (2019) courtesy of Mike Gruntman.

Fig. 6.16. Ronald Reagan in Tbilisi, Georgia. Photograph (2014) courtesy of Mike Gruntman.

6. Boundary Conditions   Mike Gruntman

Murphy added that "American Communists also were doing everything possible then [in 1940] to discourage our [U.S.] own preparedness and to oppose assistance to France and Britain."[50]

Today, many pieces of art and memorials remind us of the countless victims of Marxism. They dot the post-Communist space of the former Soviet republics and eastern European countries (Fig. 6.14). The sculptures of the communist leaders, so much celebrated by their radical Western friends and tolerated by sympathetic socialists and "balanced" liberals, moved to the museums after the decades of the dominating presence in public space (Fig. 6.15). No surprise that the lands and people liberated from the tyranny appreciate freedom and celebrate those in the free world who had the moral clarity and courage to call the Marxist regimes evil (Fig. 6.16).

Keeping the spirit of freedom alive

After watching the nightly television news program *Vremya* with featured "progressive" collaborators, or instead of watching this program, some Soviet subjects turned to shortwave radios. They tried to catch glimpses of censored information about events in the world, enjoy disapproved jazz music, and listen to the excerpts from publications by Western scholars and Soviet dissidents about forbidden topics of history, culture, and life. The Western radios fought a noble war in the ether, trying to reach listeners against formidable jamming.

The United States government established Radio Liberty and Radio Free Europe in Munich, the principal city of the American Zone of occupied Germany. The radios had their main studios and production center in buildings on Oettingenstrasse 67 in a large public park, the *Englischer Garten* (English Garden). This complex spread over almost 12 acres and "was built at initial cost of $1.1 million on property acquired under a 30-year lease in February 1951."[51] A senior RFE/RL radio executive described that this "sprawling three-floor frame construction [was] designed in the shape of a wide hand with seven stubby fingers, its palm backed up against a chain-link fence marking the boundary of the heavily wooded English Garden"[52] (Fig. 6.17).

Radio Free Europe occupied these premises from the early days of 1953. Radio Liberty was first located on the city's outskirts in a building of Munich's old pre-World War II airport, Oberwiesenfeld (Fig. 6.18), and

---

50. Murphy, 1964, pp. 34, 35
51. Comptroller General..., 1972, p. 94
52. Mickelson, 1983, pp. 5, 6

Fig. 6.17. Main production complex with studios and master control of Radio Free Europe and Radio Liberty on Oettingenstrasse 67 in the Englischer Garten in Munich, West Germany, in the 1960s. Radio Free Europe occupied these buildings since 1953. Radio Liberty joined RFE in the mid 1970s. Photograph credit: RFE/RL, USAGM.

then in the Arabellapark area of the Bogenhausen district of Munich. It had finally completed the transition to the RFE buildings in the Englischer Garten by 1976.

A government report noted in 1971 that Radio Liberty was equipped with modern studios and master control complex. It was also supported by

> a monitoring unit with the capacity to monitor up to 50 U.S.S.R. radio stations. A closed circuit television network has been installed to present monitored U.S.S.R. television programs... A library... includes 65,000 volumes and numerous subscriptions to U.S.S.R. periodicals.[53]

With so many Western intellectuals complicit in establishing and propping up Marxist regimes and ideology, the success and efficiency of foreign radios in keeping the spirit of freedom and hope alive behind the Iron Cur-

---

53. Comptroller General..., 1972, p. 119

6. Boundary Conditions   Mike Gruntman

Fig. 6.18. In the early 1950s, the American Military Government assigned this operations building of Munich's old pre-World War II airport Oberwiesenfeld to Radio Liberty. It was a historic landmark related to the infamous Munich agreement. "There in 1938 [Adolf] Hitler had greeted the British and French prime ministers when they arrived for the conference that sealed Czechoslovakia's fate" (Critchlow, 1995, p. 3). Later, the local authorities razed the building during the development of the area for the Olympic summer games of 1972. Photograph credit: RFE/RL, USAGM.

tain were truly remarkable. The members of the Politburo Ligachev and Chebrikov provided an overview of the state of foreign broadcasts in 1986:

> Broadcasts by nongovernmental radio stations "Radio Liberty," "Radio Free Europe" as well as such radio stations as "German Wave" and "Voice of Israel" have openly anti-Soviet character and are full of spiteful slanders about the Soviet realities.
>
> Radio stations "Voice of America" and "BBC" give their materials usually tendentiously, from the anti-Soviet perspective but try at the same time maintain an objective approach to events and facts of international life, politics, economy, and culture.[54]

---

54. Ligachev and Chebrikov, 1986

These two senior Soviet leaders continued then disparaging the quality of broadcasts of their ideological siblings and political competitors in the Marxist world in the People's Republic of China and North Korea. Ligachev and Chebrikov proposed concentrating on jamming foreign radios which presented, in the view of CPSU, the main threat to the Soviet regime and expanding communist propaganda in the free world:

> Propaganda materials of radio stations "Radio Peking [Beijing]" and "Radio Korea" are unfriendly towards the USSR but unconvincing and their presentation is biased and primitive.
> ...
> Under such conditions, it is advisable to stop jamming government radio stations "Voice of America" and "BBC" as well as "Radio Peking" and "Radio Korea."
> The released technical capabilities are to be used for higher quality and more reliable jamming of radio stations "Radio Liberty," "Radio Free Europe," "German Wave," and "Voice of Israel." It is advisable to [also] use part of the [released] technical capabilities to expand Soviet broadcasting to capitalist countries.[55]

Clearly, Politburo viewed Radio Free Europe and Radio Liberty, together with Voice of Israel and German Wave, to be particularly effective in the ether.

Such assessment had deadly consequences on the ground from the early days of the Radios. The communist intelligence services continuously tried to sabotage RL and RFE. They killed two Radio Liberty staff members, blackmailed many others, and attempted to poison Radio Free Europe employees in the 1950s. Intelligence officers infiltrated the radio staff and organized a bomb explosion (Fig. 6.19) at the main RFE/RL production and office center in Munich's Englischer Garten in February 1981.[56]

To my delight, the Radio Liberty officials arranged for me to visit their headquarters when I was in Prague for a few days in 2008. I wanted to express sincere thanks to the organization that did so much to keep the flame of free thought alive in totalitarian hell. By that time, the RFE/RL operations had moved from the original premises in Munich to Prague in the Czech Republic. In addition to a tour of the studios, the communications office set up a friendly meeting over coffee with two editors of the Russian-language service, Ivan Tolstoi and Victor Yasmann[57] (Fig. 6.20).

---

55. Ligachev and Chebrikov, 1986
56. e.g., Frolik, 1975, pp. 38, 39; 1976, pp. 49, 50; Meyer, 1980, p. 120; Mickelson, 1983, pp. 8, 9; Pacepa, 1987; Critchlow, 1995, pp. 55-60; Sosin, 1999, pp. 192-193; Kalugin, 2009, pp. 224, 225;
57. Yasmann, n.d.

6. Boundary Conditions   Mike Gruntman

Fig. 6.19. Damaged headquarters of Radio Free Europe and Radio Liberty in Munich, West Germany, by a bomb explosion in February 1981. Photographs credit: RFE/RL, USAGM.

Fig. 6.20. Author (left) with Radio Liberty editors Ivan Tolstoi (right) and Victor Yasmann (center) in a cafeteria of the RFE/RL headquarters in Prague, the Czech Republic, on April 22, 2008. Photograph from collection of Mike Gruntman.

We quickly found somebody whom we all knew. Almost twenty years earlier in 1989, the Academy of Sciences sent Baranov and me to West Germany for several days as part of our heliospheric collaborative effort. While in Munich during the weekend, I telephoned Radio Liberty to pass a message, "feedback" about coverage of their local events, from my friends in the contested Abkhazia region in the Caucasus. I talked then to one of the editors, who was a former defector from the Soviet Union. He happened to be an old colleague and friend of one of my Radio Liberty hosts in Prague. The latter told me interesting details about his friend. As a KGB officer, he was sent to accompany a Soviet tourist group with the task of watching and preventing the defection of the tourists, when he decided not to go back himself.

My telephone conversation with Radio Liberty in Munich took place on July 16, 1989. By coincidence, on that same day, major riots and interethnic fighting erupted in Sukhumi, the capital of Abkhazia, with more than a dozen people killed and hundreds injured. So the feedback from my friends was very timely and much appreciated by the radio editor. He invited me to come to his office for a chat. "You just need to show your passport at the entrance. We do not record the names," he said. I declined, telling him that they were likely penetrated by the KGB. He did not disagree with the common-sense response.

Printed word

In selected kiosks in downtown Moscow, one could buy foreign newspapers. Only communist ones, of course. Even such a marginal publication as *Morning Star*, the organ of the small British Communist Party, sometimes printed factual information purged from the Soviet press. This Kremlin lap dog party was, in an assessment by a knowledgeable Czechoslovakian intelligence officer, a "downright failure" politically except for its role in influencing British trade unions.[58]

The French *L'Humanité* did not follow the Moscow line on some issues in those days and sometimes had interesting articles. This newspaper represented a very large and influential French Communist Party, receiving one-fifth of the votes in national elections. Similar to France, powerful communist parties in Spain and Italy also played with the "heresy" of Eurocommunism, periodically upsetting their Moscow patrons.

A major national library in Moscow, the *All-Union Library of Foreign Languages* (Fig. 6.21), had a large collection of Western publications. Its books covered specialized topics. They typically presented no ideological threat to the regime and did not interest the general public. The library primarily served specialists in foreign languages, philology, philosophy, literature, history, and other non-hard-science areas. Access to the library required proof of being a student or a "legitimate" specialist such as, for example, a research fellow in the Academy of Sciences.

The only daughter of Chairman of the USSR Council of Ministers Aleksei Kosygin, Lyudmila Gvishiani-Kosygina, served as the library director. Her husband, Dzhermen Gvishiani, was also a high-level Soviet official. Perhaps due to the connections, the library had adequate resources and friendly staff who were not overworked. It kept doors open until late in the evenings and during weekends. This comfortable place had a clean cafeteria.

The library collections included some publications, with access usually

---

58. Frolik, 1976, pp. 111, 112

Fig. 6.21. Library of Foreign Languages in the central part of Moscow in June 2016. Building's appearance did not change since the 1970s and 1980s. Photograph courtesy of Mike Gruntman.

restricted to ideologically "anointed" Communist Party officials and the like. Therefore, one could find there books that did not exactly align with the "only scientific and accurate" Marxist-Leninist version of history and the view of the world propagated by the communists. It even had a few scholarly textbooks of Hebrew. At that time the authorities harshly suppressed attempts to learn and especially teach this language.

For me, the library became a window to a bigger world. Besides, it helped improve my English. My teacher in the secondary school and instructor in Fiztekh had never visited an English-speaking country or perhaps any other place abroad. Not only that, their English teachers and likely the teachers of their teachers had also never set foot beyond the Soviet state borders.

A major treatise by Winston Churchill, "Second World War," became my favorite. This monumental six-volume work mightily contributed to the Nobel Prize in literature bestowed on him in 1953. I read Churchill's thick volumes on my periodic visits to the library for several years. They were always available to check out for reading in the library halls. This availability indicated that somebody else very rarely requested them, justifying lax restrictions on book access.

6. Boundary Conditions   Mike Gruntman

Fig. 6.22. The author of this book meets his teacher in the world of policies, geopolitics, and international relations on a London street in March 2002 (left) and in Paris in 2010. Photographs from collection of Mike Gruntman.

Churchill's volumes became my education in the fields of political science and international relations. One did not have to agree with his positions, the flavor of his descriptions, or his conclusions. Sir Winston was a lion guarding, roaring, and ferociously fighting for the interests of *his* empire. It is not entirely without justification that the French called their perennial British rivals la *perfide Albion*, or the perfidious Albion.

At the same time Churchill's superb analyses, broad horizons, and independent thinking identified him as a unique and great statesman and forceful leader. The United Kingdom of his days was still an empire, although in retreat. Consequently, Churchill's writings taught about the development and implementation of policies and politics on a global scale, covering the entire world. His treatise opened unmatched vistas for seeing and understanding the complexities of multi-faceted relations among the nations, friends, and foes. Sir Winston became my true teacher in these areas (Fig. 6.22).

The lessons of the world-war events in the twentieth century, which Churchill called "Moral of the Work," have been guiding me since those days: "*In War:* Resolution. *In Defeat:* Defiance. *In Victory:* Magnanimity. *In Peace:* Goodwill."

One often hears embarrassingly primitive nonsense from academics, journalists, and assorted pundits on television and in the press, all university graduates. In my view, reading (and supplementing with homework of thinking through) Churchill's entire fundamental treatise "Second World War" should earn one a bachelor's or master's degree in political science and international relations.

Under these boundary conditions of life and work, our development of position-sensitive detectors had shown major successes by 1983. Its advance continued gradually and slowly afterward, despite enthusiasm, energy, and expertise, as well as interest by many research organizations. Concurrently, studies of neutral atoms in space and the development of experimental techniques to explore the neutral solar wind and heliospheric energetic neutral atoms from the interstellar boundary of the solar system began taking more and more time and effort.

7. Energetic Neutral Atoms

Birth of ENAs

Local, *in situ*, direct detection of individual neutral atoms in interplanetary space looked unrealistic for decades. Mass spectrometers have been measuring neutral atoms and molecules in the upper atmosphere of the Earth since the 1950s. Such instrumentation could probe, in situ, relatively large number densities in planetary environments. Optical sensors also detected solar ultraviolet and extreme ultraviolet photons scattered by atoms in interplanetary space. These observations were, however, essentially non-local, line-of-sight, and limited in sensitivity.

Charged particles (ions) and neutral particles collide in space plasmas. Then, an exchange of charge sometimes occurs when an electron jumps from one particle to another. The process is known as *charge exchange*. Whenever an energetic ion undergoes a charge exchange collision with a neutral background atom, the ion becomes an energetic neutral atom (Fig. 7.1). Planetary atmospheres and interstellar gas that fills interplanetary space provide neutral atoms for energetic ions to become neutral. Consequently, such collisions create ENAs everywhere. The number densities of ENAs are exceptionally low and their fluxes are usually weak.

Magnetic fields permeate space, whether near planets (forming magnetospheres) or in interplanetary and interstellar space. Energetic ions cannot travel across magnetic field lines. The field bends ion trajectories by the Lorentz force and causes them to gyrate (Fig. 7.2). Therefore, the magnetic field confines the ions and prevents them from reaching a remote observer. Consequently, one cannot directly examine the properties of ions in, for example, a magnetosphere from a distance.

7. Energetic Neutral Atoms — Mike Gruntman

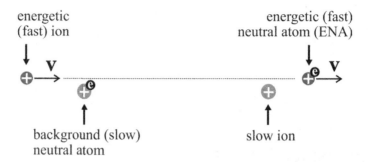

Fig. 7.1. Charge exchange collision of an energetic (fast) ion and a background (slow) neutral atom. A negatively charged electron **e** jumps from one colliding particle (neutral atom) to another (ion). The created neutralized ion, the energetic neutral atom or ENA, preserves the magnitude and direction of the original velocity vector, **V**, of the energetic ion.

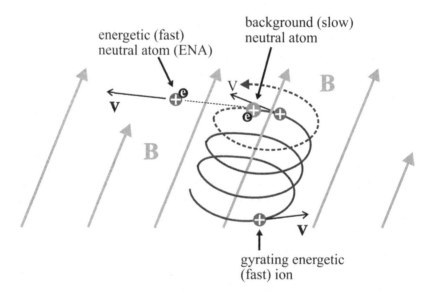

Fig. 7.2. Energetic ions gyrate in magnetic field **B** in space. After a charge exchange collision, the newly created ENA preserves the velocity vector **V** of the ion at the moment of the collision. It then flies away across magnetic field lines in a straight trajectory as a stone from a sling. Consequently, one can probe the properties of energetic ions remotely from a large distance by detecting ENAs. Recording an image of a plasma object, such as a magnetosphere or the solar system interstellar boundary, formed by emitted energetic neutral atoms became known as ENA imaging.

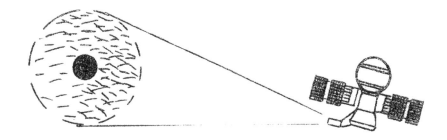

Fig. 7.3. An instrument onboard a spacecraft performing magnetosphere imaging in ENAs. Figure from Gruntman and Leonas, 1986.

In contrast to ions, a unique feature of ENAs is that magnetic fields do not exert forces on neutral atoms. Therefore, the newly born ENAs fly across magnetic field lines along straight trajectories (Fig. 7.2) until lost due to ionization. They could cover large distances in space. The gravitational field of the Sun and planets may bend ENA trajectories, but this effect is often small and can be accounted for.

An observer can thus measure ENA fluxes produced by ions in remote plasmas at large distances. Obtaining an image of a space plasma object such as a planetary magnetosphere (Fig. 7.3) or the interstellar boundary of the solar system in emitted ENA fluxes became known as *ENA imaging*. Scientifically inclined readers can find physics details of the concept and instrumentation in a review article in a physics journal.[1]

ENA imaging of space plasmas is conceptually similar to widely used passive corpuscular diagnostics of fusion plasmas.[2] Already in 1951, Igor E. Tamm[3] noted that a hot dense plasma with thermonuclear reactions would be a source of abundant energetic neutral atoms created in charge exchange. His first estimates showed that such atoms would account for most of the energy transfer to the reactor wall—much more than by plasma thermal conductivity.[4] He did not explicitly mention the possibility of measuring this flux of energetic neutral atoms for probing plasma properties, but this would have been the next logical step.

1. Gruntman, 1997a
2. e.g., Eubank, 1979; Medley et al., 2008; Petrov et al., 2021
3. Igor Tamm, Pavel Cherenkov, and Ilya Frank received the Nobel Prize in physics in 1958 for the discovery (in 1934) of what would become known as *Cherenkov radiation*.
4. Tamm, 1958 (English translation 1961)

While similar to ENA imaging in space, passive corpuscular diagnostics of fusion plasmas also exhibit essential differences. The short-lived fusion plasmas typically produce high fluxes of neutral atoms. A relatively large plasma density makes it not "transparent" to ENAs with energies below 10-20 keV, which are of primary interest in space. The latter feature limits the technique to probing the edges of dense fusion plasmas. In space, tenuous plasma objects are usually "ENA-thin" and ENA fluxes are very weak.

Wandering in wilderness

The presence of energetic ions and neutral gas everywhere in space makes ENAs ubiquitous. The ideas of probing planetary magnetospheres and the interstellar boundary of the solar system in energetic neutral atoms originated in the early days of the space age. Scientific literature occasionally mentioned, as a curiosity, the possibility of the existence of charge-exchange-produced neutral atoms with energies of hundreds and even thousands of electronvolts.

As early as 1961, Dessler, Hanson, and Parker noted that observing fast hydrogen atoms emitted by a magnetosphere could serve as a tool for studying proton ring currents in geomagnetic storms.[5] These ENAs in today's terminology had to be produced in charge exchange between the magnetic storm protons and exospheric hydrogen atoms. Then, two years later, Patterson, Johnson, and Hanson suggested[6] that a significant fraction of distant solar-wind protons could reenter the heliosphere in the form of hydrogen ENAs as a result of processes at the interstellar boundary of the solar system. At that time, Akasofu also advanced a hypothesis of large numbers of energetic hydrogen atoms present in the solar wind as transients during violent events.[7]

Energetic neutral atoms open a way to probe remotely global ion and plasma properties. From a distance, one could obtain an image of a distribution of energetic ions surrounding a planet, its magnetosphere, by detecting and identifying individual energetic neutral atoms originating there. Figure 7.3 shows a spacecraft capturing a series of "enagraphs," as envisioned by us in the mid-1980s, which would reveal "the time evolution of the magnetospheric processes, for instance of the ring current decay."[8] The same publication also described in detail for the first time a potential promise of coded apertures for ENA imaging. Several years later, when I was

---

5. Dessler et al., 1961
6. Patterson et al., 1963
7. Akasofu, 1964
8. Gruntman and Leonas, 1986, pp. 3, 4

already at USC, it became clear that this powerful experimental technique would be unsuitable for magnetosphere imaging.[9]

One could also map the interstellar boundary of the solar system 100 AU away from the Sun, where the expanding solar wind rams into the interstellar wind, a stream of partially ionized interstellar gas. Proton-neutral collisions at this remote region would produce heliospheric ENAs. In the early 1980s, we emphasized that

> detecting the flux of such [energetic neutral] atoms [from the interstellar boundary], that could be called the "reflected solar wind," would be the first [direct] experimental source of information about the processes in the region of [the] solar wind termination,[10]

with such an experiment promising "unique scientific" results. The idea of detecting ENA messengers from the interstellar boundary of the solar system looked fantastic in those days, truly out of this world, and has motivated much of my work since the late 1970s.

The introduced terms "enagraph" and "reflected solar wind" did not get traction and are not used today.

In the 1970s, several groups in the world optically studied interstellar hydrogen and helium atoms that entered the heliosphere by measuring solar light scattered by them. In IKI's Department No. 18, Vladimir Baranov theoretically explored global plasma flows in the solar wind interaction with the local interstellar medium. In 1971, he formulated a concept of two colliding supersonic flows of the solar and interstellar winds. A few scientists in his theoretical group took part in this work, at first Mikhail (Misha) Ruderman in the 1970s and later Yuri Malama and Sergei Chalov (Fig. 7.4). Vlad Izmodenov joined the work in the 1990s.

In contrast to some interest in plasma flows and interstellar gas, no attention was paid to energetic neutral atoms in those early days because nobody knew how to detect such atoms. On the other hand, estimates of expected ENA intensities and energies were essential from a practical point of view to guide the development of the instrumentation. Consequently, the calculations of interstellar helium fluxes first appeared in my master's thesis (Fig. 3.5) in the mid-1970s and then expanded to other ENA fluxes.

With time, more scientists became independently interested in magnetospheric ENAs, notably Ed Roelof at the Applied Physics Laboratory and Hank Voss at *Lockheed-Martin* in the United States in the early 1980s. There had been even attempts, with inconclusive results, to measure the precipitat-

---

9. Gruntman, 1993c
10. Gruntman and Leonas, 1983, pp. 25, 26

Fig. 7.4. Scientists in Vladimir Baranov's theoretical laboratory in Department No. 18 of IKI (and later at IPM) studying global interactions of the local interstellar medium with the solar system.

Top: Mikhail (Misha) Ruderman (in 2018); middle: Yuri Malama (in the mid-1990s); bottom: Sergei Chalov on the left and Vlad Izmodenov on the right (in 2005).

Izmodenov got involved in the heliospheric work of Baranov as his student in the 1990s. Today, he heads a group at IKI that leads heliospheric studies in Moscow. Vlad is also a professor at the Moscow State University.

Photographs courtesy of Mike Gruntman.

ing magnetospheric ENAs a decade earlier. Bill Bernstein and coworkers at *TRW Systems* in Redondo Beach, California, attempted such measurements using the first-ever dedicated ENA instruments on sounding rockets in 1968 and 1970. TRW initially supported these experiments through its internal independent research program, with NASA subsequently stepping in.

Oleg Vaisberg's group at IKI included a separate channel in the ion instrument RIEP that could detect, as they apparently hoped and believed, high-intensity neutral atom fluxes in the solar wind on the Soviet *Mars-3* interplanetary mission in 1971. The attempt did not succeed.

Much less was known about ENAs from the interstellar boundary of the solar system. The interaction of our star, the sun, with the surrounding interstellar medium forms the heliosphere, a region where the Sun controls the state and behavior of the plasma environment (Fig. 7.5). The heliosphere is a complicated phenomenon where the solar wind and interstellar plasmas, interstellar neutral gas, neutral solar wind, magnetic field, and cosmic rays play prominent dynamic roles.

It had been clear since the late 1970s that in order to establish the feasibility of mapping the interstellar boundary in ENA fluxes, one needed "accurate estimates of the expected differential (in energy and direction) flux densities" of heliospheric ENAs and that "the existing estimates could not be considered to be complete and reliable."[11] Theoreticians ignored this entire area.

It would take years to firm up the concept of ENA imaging of the heliosphere after the first steps of the 1970s. A consequential advancement happened in 1992. Already at USC, I showed that expected heliospheric ENA fluxes from the interstellar boundary reaching 1 AU from the Sun should be highly anisotropic.[12] This feature logically pointed to the promise of measuring the directional dependences of ENA intensities, which would become known as ENA imaging of the heliosphere. The all-sky map in fluxes of energetic neutral atoms must be sensitive to the structure and processes taking place 100 AU away from the Sun. The heliospheric ENAs would serve as trusted messengers of the conditions at the solar system frontier.

The road to realizing a concept in a space mission is long. It takes the effort of many scientists and engineers, indispensable institutional support, and significant government resources. Our work in the 1970s, 1980s, and early 1990s contributed to building up the necessary foundations for future ENA imaging missions to study the heliosphere and planetary magnetospheres.

---

11. Gruntman and Leonas, 1983
12. Gruntman, 1992a

7. Energetic Neutral Atoms — Mike Gruntman

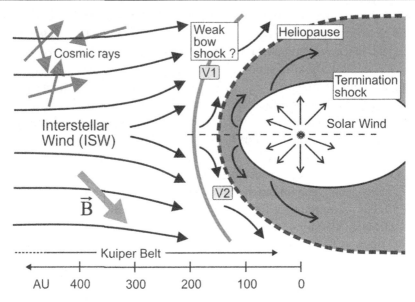

Fig. 7.5. The Sun's heliosphere and its galactic neighborhood. A supersonic radial expansion (with a velocity of 500 km/s) of the solar wind ends with a termination shock where the solar wind plasma abruptly slows down and heats up. The heliopause (dashed line) separates the solar and interstellar plasmas and forms the boundary of the heliosphere. The flow of interstellar plasma and neutral gas with a velocity of 25 km/s, or 5 AU/yr, relative to the Sun, the interstellar wind, may form a weak bow shock in front of the heliosphere. The surrounding local interstellar medium, LISM, includes cosmic rays and carries the frozen-in magnetic field **B**. Interstellar helium atoms fly into the solar system practically unimpeded and reach Earth orbit, where they can be directly detected.

The figure also shows the positions of Voyager 1 (V1) and 2 (V2) spacecraft at the heliocentric distances of 165 AU and 138 AU, respectively, in 2025. These space vehicles fly away from the Sun. Voyager 1 and 2 crossed the termination shock in 2004 and 2007 and the heliopause in 2012 and 2018, respectively.

The gray region between the termination shock and heliopause, the heliosheath, contains the solar wind plasma heated in the shock transition. Charge exchange collisions between hot plasma protons and inflowing interstellar hydrogen atoms produce heliospheric energetic neutral atoms that can reach our planet Earth.

Measuring the directional intensities and energy dependencies of these heliospheric ENAs on a spacecraft near the Earth remotely probes the heliosheath and maps the interstellar boundary of the solar system. Development of the experimental concept and instrumentation for ENA imaging of the heliosphere drove our work at IKI starting in the late 1970s. NASA's IBEX space mission mapped the interstellar boundary in ENA fluxes for the first time 30 years later in 2009 (Fig. 3.8).

Detecting neutral atoms directly

When a neutral atom impacts a surface, it may knock out an electron and/or ion. It may also bounce off the surface as an ion. Then one could pick up this emitted or reflected charged particle, analyze it, and register it with a secondary electron multiplier.

If one put an open secondary electron multiplier on a spacecraft, it would detect ENAs directly. However, the noise count rate due to background ultraviolet photons would be six to eight orders of magnitude higher. Suppression of this gigantic background in a realistic experiment looked hopeless in the 1970s. Nevertheless, I jumped on the challenge of ENA detection in space. This meant finding physical phenomena caused by ENAs in sensors that could be unambiguously and reliably detected in the presence of a superior flux of background photons.

The expected energies of interstellar helium atoms ranged from 50-150 eV at 1 AU (Fig. 3.5). I was not particularly optimistic about their direct detection and gradually turned attention to a more realistic registration of ENAs with energies of a few hundred and perhaps one thousand electronvolts. In contrast to interstellar helium atoms, hydrogen atoms with such energies could also pass through ultrathin foils, the phenomenon enabling their detection.

One could identify three populations of such higher-energy atoms in space. First, the ENAs with energies 1-2 keV should exist in the solar wind. This is the neutral solar wind. Second, a fraction of solar wind protons would charge exchange at the interstellar boundary of the solar system and return back to the inner solar system as heliospheric ENAs. Third, a planetary magnetosphere could also be a source of ENAs. Radiation belts consist of energetic ions and exospheres extend to large distances, providing abundant neutral atoms for charge exchange. Therefore, charge exchange collisions should occur in magnetospheres of planets, producing ENAs with energies as high as a few dozen kiloelectronvolts.

All three types of energetic neutral atoms had energies sufficient for knocking out electrons from surfaces and penetrating through ultrathin foils. Consequently, the future instrumentation for these ENAs would share common approaches.

There was also another important family of non-thermal, energetic neutral atoms with energies under 40 eV. Such energies were too low for penetrating the foils and for efficient secondary emissions. These atoms were primarily represented by interstellar hydrogen and traces of deuterium flying into the solar system from the LISM and filling interplanetary space.

7. Energetic Neutral Atoms Mike Gruntman

Interstellar oxygen atoms also fell into this category of particles with similar energies per nucleon and with large electron affinities. The latter meant that they could form negative ions. (Note that negative ions of helium cannot exist in a stable form.) Labeled as "low-energy ENAs," these atoms required physically different approaches for detection. Atoms with such energies per nucleon should also originate in planetary ionospheres and one would encounter them in cometary flybys.

Figure 7.6 shows several basic common techniques and elements of present-day ENA instrumentation. Free-standing ultrathin carbon foils (Fig. 7.6,a), only 10-40 atomic layers thick, had been used in nuclear physics experiments for some time by the late 1970s. Atomic particles with energies of several hundred electronvolts per nucleon and higher could penetrate and fly through them. The first ENA-detection experiments on sounding rockets in 1968 and 1970 relied on stripping neutral atoms (converting to positive ions) passing through such foils (7.6,b), with the subsequent ion analysis and detection. IBEX would later use this approach for mapping the interstellar boundary in ENA fluxes. Other techniques shown in Fig. 7.6 had been used in unrelated applications under different environments and for different energies, species, and wavelengths in nuclear physics experiments (7.6,c), production of high-intensity negative ion beams (7.6,d), and in "superinsulator" shields in the infrared spectral region (7.6,e).[13]

It was important that ENA instruments also rely on microchannel plate sensors, including those with position sensitivity, for the detection of ions and electrons. Consequently, my ongoing development of MCP detectors and PSDs nicely complemented and contributed to the effort in ENA imaging.

During my 15 years at IKI, we advanced, improved, developed, proposed, or contributed to the advancement of all techniques shown in Fig. 7.6 or their elements and components.

In modern-day science, the knowledge and expertise are highly specialized, which makes work in unrelated fields difficult. It is very hard to concurrently design and build vacuum chambers, produce well-characterized neutral beams, develop new and unconventional detectors and sensors, and theoretically explore (including by computer simulations) space physics phenomena. The reality of the ENA detection challenge required working in all those fields, which made professional life certainly not boring during all my time at IKI. As described below (Chapter 8), helpful external collaborations, primarily foreign, would emerge with time and enable progress.

---

13. Gruntman, 1997a

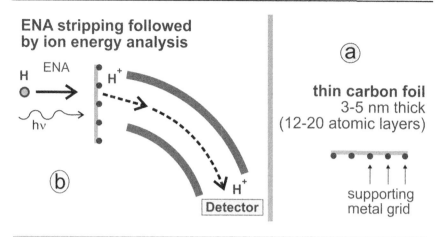

Fig. 7.6. Basic techniques and sensor components for ENA detection. (a) – ultra-thin carbon foils for detection of ENAs with energies >400 eV. (b) – ENA stripping in thin foils with subsequent ion analysis. (c) – thin-foil-based time-of-flight ENA detection. (d) – detection of low energy (<40 eV/nucleon) ENAs by converting them into negative ions (left) or sputtering positive and/or negative ions from the surface (right). (e) – filters with small holes or slits allowing passage of ENAs and blocking background EUV and UV photons.

1979 List

Possible approaches to detect energetic neutral atoms while suppressing background photons had been identified by 1979. That year, I compiled a list of a dozen ideas for ENA detection on one sheet of paper. This handwritten *1979 List* became my lodestar. It included three "top-level" main tasks or concepts.

First, to lower the energy threshold by 1-2 orders of magnitude in thin-foil-based time-of-flight particle detectors of the type used in nuclear physics experiments (Fig. 7.6,c). Second, to develop detection techniques based on the conversion of incident ENAs on surfaces into negative ions (Fig. 7.6,d, left). And third, to explore the feasibility of filters based on holes. Such structures should allow the passage of neutral atoms while blocking background ultraviolet photons (Fig. 7.6,e). The filters could dramatically improve the ratio of the signal particles (ENAs) to noise background photons entering instruments.

All three "raw" main ideas from the 1979 List required serious experimental development. We advanced them to various degrees during my time at IKI, and all three, with time, would prove successful in ENA instrumentation in space missions 20-30 years later.

In addition, the list included several "supporting" items to explore which could make ENA detection more reliable. In particular, I identified scattering in foils, triple coincidences, and amplitude selection of MCP pulses.[14] The latter idea would motivate our later work on statistics of electron emission from foils caused by passing particles and require the development of MCP assemblies with record amplitude resolutions.

Reduction of energy threshold

By 1981 we had experimentally demonstrated the first top-level item in the 1979 List, the reduction of the energy threshold for detection of ENAs, to a few hundred electronvolts.[15] At that time, nuclear-physics-type thin-foil-based sensors detected particles with energies 20 keV and higher. It took a couple of years to build and optimize the required MCP sensors, explore the properties of ultrathin foils, and develop the nanosecond time-of-flight electronics.

Some of our ENA analyzers relied on two MCP detectors operating in the two-channel TOF mode (Fig. 7.6,c). In addition, we introduced a compact time-of-flight sensor that used only one MCP detector. This latter concept had already emerged in 1977-1978 when the focus of my work

---
14. Gruntman and Leonas, 1983
15. Gruntman and Morozov, 1981, 1982; Gruntman, 1983

began turning toward the detection of energetic hydrogen atoms of the neutral solar wind.

Our ENA analyzers with one MCP detector operated with either a single collector (anode) in the single-channel TOF mode or a divided collector in the conventional two-channel TOF mode. ENA instruments on NASA's IMAGE (MENA) and TWINS missions would use the divided collector designs.[16] These instruments became operational in 2000 (IMAGE), 2006 (TWINS-1), and 2008 (TWINS-2). (IMAGE stood for *Imager for Magnetopause-to-Aurora Global Exploration*; MENA for *Medium Energy Neutral Atom Imager*; and TWINS for *Two Wide-Angle Imaging Neutral-Atom Spectrometers*.)

To the best of my knowledge, nobody has ever used a particle analyzer with a single-channel time-of-flight mode in space.

This early work led to another new promising technique for measuring "absolute intensities of neutral or charged particle fluxes without preliminary detector calibration, provided that there is no background."[17] The approach took advantage of coincidences of physically independent events produced by the same incident particle.

As I found many years later, Johannes "Hans" Geiger and Otto Werner had demonstrated this feature of coincidences in the 1920s when they determined the "detection efficiencies" of human observers registering light flashes on phosphor screens.[18] The IBEX mission applied this technique for the performance evaluation of the ENA IBEX-Hi instrument.[19]

Two other top-level ideas from the 1979 List took much longer to develop.

Surface conversion to negative ions

Low-energy ENAs with energies in the 1-40 eV/nucleon range required a physically different approach for their detection. Advances in generating high-intensity high-energy neutral atom beams for fusion plasma energy pumping and futuristic space weapons suggested a possible technology: surface conversion to negative ions. There were also independent proposals, unrealized at that time, to use this physical effect for corpuscular diagnostics of fusion plasmas. The surface conversion approach particularly suited the direct detection of interstellar hydrogen, deuterium, and oxygen atoms entering the solar system from the LISM.[20]

---

16. Pollock et al., 2000; McComas et al., 2009a
17. Gruntman and Morozov, 1981, 1982
18. Rutherford et al., 1930, p. 548
19. Funsten et al., 2005
20. Gruntman, 1993a,b; Gruntman, 1997a

7. Energetic Neutral Atoms — Mike Gruntman

I could not explore atom conversion to negative ions in the laboratory in the 1970s and 1980s. While it was clear how to do it, there were simply no funds, support, time, and other indispensable resources to build a dedicated experimental setup with an ultra-high clean vacuum and a well-characterized source of collimated monoenergetic low-energy neutral atoms. I could only follow the physics literature on the subject. We briefly described the promise of this detection technique in a publication in 1983.[21]

After reaching California in March 1990, I outlined this enabling concept first in USC technical reports in October 1990 and February 1991.[22] In August 1991, I led a team as a principal investigator (with coinvestigators from USC, University of Arizona, Stevens Institute of Technology, and Jet Propulsion Laboratory) that submitted a comprehensive proposal (nearly $800k in current dollars) to NASA: "In situ measurement of low-intensity and low-energy neutral atom fluxes in the solar system." The concept called for an in-depth experimental study of the underlying physics.

We presented the surface conversion technique, based on the proposal, at the annual Fall Meeting of the *American Geophysical Union* (AGU) in December 1991 and then at the *World Space Congress* in Washington, DC, in the summer of 1992.[23] The proposal to NASA was not funded.

In the mid-1991, the editor of the *Journal of Geophysical Research—Space Physics*, or JGR, Christoph Goertz, agreed to accept for consideration my manuscript with a detailed feasibility study of this novel instrumentation concept, although JGR at that time avoided instrumentation-focused topics. Only one week after our last telephone conversation, a disgruntled student shot Chris dead at the University of Iowa campus in a well-known tragic event on November 1, 1991. Jim Van Allen stepped in temporarily as the journal editor and confirmed Chris's commitment.

The article manuscript was ready in a few months. Jim told me that he would be stepping down soon and asked to discuss the arrangement with the new forthcoming editor. The latter declined to consider the feasibility-study-type article for publication. From the spring of 1992, I circulated the original manuscript intended for JGR among several colleagues, particularly those working on a Small Explorer proposal "HI-LITE," who were interested in the detection of neutral atom fluxes.[24]

After some delay, this article[25] describing in detail the surface conver-

---

21. Gruntman and Leonas, 1983
22. Gruntman, 1990, 1991a
23. Gruntman et al., 1991, Gruntman, 1992b
24. Smith et al., 1993
25. Gruntman 1993a

sion technique appeared in another journal, *Planetary and Space Science*, in 1993. Then *Advances in Space Research* also published another[26] of my articles that was based on a talk on the same subject given earlier at the World Space Congress in 1992.

Nobody has a monopoly on ideas for solutions to outstanding problems. After learning about my work on conversion to negative ions, a colleague at the University of Bern, Peter Wurz, wrote me in January 1993 that because of his "background in surface science" he had independently identified sometime in 1992 negative surface ionization as "the method of choice" for direct detection of low-energy neutrals.[27] It was natural that other scientists came to similar ideas, indirectly validating the new concept. The time for the technique had arrived.

My follow-on attempts to obtain funding for an experimental program to implement ENA surface conversion to negative ions did not succeed. Others would build the space instruments. Nevertheless, I was glad to publish a detailed feasibility study that helped others advance the method and avoid, as I hoped, missteps in its development and implementation. Space instruments based on the surface conversion to negative ions, first conceived at IKI in the late 1970s, flew later on NASA's space missions IMAGE (LENA) and IBEX (IBEX-Lo) and are planned for future experiments.

Filters to improve the signal-to-noise ratio

The separation of ENA particles from accompanying intense EUV and UV background radiation is one of the most important requirements for ENA instruments.

The idea behind the filters, the third-highest priority task in my 1979 List, is very simple. It relies on photons' ability to pass through a straight channel-pore (or a slit) in a filter film (or plate) only if its channel diameter (slit width) is much larger than the photon wavelength. When the size of openings is not much larger than the wavelength, then diffraction effects occur at the entrance and exit of the channel (slit) and the channel itself serves as a waveguide attenuating the electromagnetic radiation.

In contrast to photons, an ENA follows a straight trajectory and passes through a channel freely if it does not collide with the walls. The De Broglie wavelength of an ENA is tiny compared to realistic openings. Such filters, sometimes loosely called *diffraction filters* in the early days, could thus separate incident ENAs from background EUV and UV photons and simultaneously serve as particle collimators.

---

26. Gruntman 1993b
27. Wurz, 1993

Holes had been used to control radiation transmission in the far-infrared spectral band for some time. Metal meshes also served as mirrors, efficiently reflecting radio waves in antennas. Abundant background energetic photons in space are primarily at resonance transitions in atoms of hydrogen (121.6 nm) and helium (58.4 nm). Therefore, their suppression required filters with sizes of openings 100 nm (1000 A) or smaller.

Looking for ways of implementing this idea, I learned in 1979 about the so-called nuclear track filters or NTFs. A few groups fabricated such filters at particle accelerators across the world by etching nuclear tracks produced by the penetration of thin (1–20 µ) films by high-energy (1 MeV/nucleon) heavy (>100 amu) ions. This fascinating nuclear-physics technology reliably produced almost perfect cylindrical pores with diameters varying from 4–10000 nm (0.004–10 µ) in films. Diverse applications of NTFs included diffusion enrichment of uranium, separation of cancer cells in the blood, and clarification and cool stabilization of wine and beer by sieving out bacteria, sediments, and yeast.

In the late 1970s, a group at a heavy-ion accelerator in Darmstadt, West Germany, led the development of nuclear track filters. In the USSR, scientists working for academician Georgii N. Flerov[28] made such NTFs at a high-energy particle accelerator at the *Joint Institute for Nuclear Research* in Dubna.

Nuclear physicist Flerov was an influential figure in the scientific establishment. While serving in the armed forces during World War II, Flerov wrote a letter to Stalin in 1942. He pointed out the abrupt disappearance of publications on nuclear fission in foreign scientific publications and urged the initiation of a national program to build an atomic bomb. The government transferred Flerov to the Academy of Sciences the same year where he soon joined the fledgling atomic weapons program. A superheavy artificial element, Flerovium, with the atomic number 114, was named after him.

One day Leonas and I took a suburban train for a three-hour ride to Dubna. There, we told Flerov about the idea of filtering out extreme ultraviolet photons in space. He liked the concept and gave the green light to his "boys" to help us with filters. No contracts or formal agreements, just a scientist to a scientist, one group to another.

It so happened that, at that time, I was giving advice on microchannel plates and position-sensitive detectors to a group in the *Physical Institute of the Academy of Sciences*, or FIAN. One of their scientists, Alexander Mitrofanov, was very much interested in nuclear track filters for imaging the Sun

---

28. sometimes spelled Flyorov

in extreme ultraviolet and soft X-rays. Importantly, they had experimental capabilities to measure filter transmissions in the desired spectral bands.

We quickly teamed up. Mitrofanov experimentally proved efficient suppression of the incident EUV and UV radiation by NTFs and the filters looked promising for the intended applications of the FIAN group. They did not suit, however, ENA imaging because of fundamental limits of geometric transparency.

The etched nuclear-track holes were randomly distributed across filter surfaces (Fig. 7.7). To preserve the structural integrity of self-supporting filters under the severe dynamical conditions (shocks, vibrations) of a space launch, the number of holes had to be limited. This, in turn, required the total average geometric transmission of filters not to exceed 1% or, preferably, 0.1%. So, while an NTF could significantly improve the ratio of ENAs to background photons entering a sensor, such a filter would have reduced the already low incident ENA fluxes by an unacceptable factor of one thousand. Consequently, we had to abandon the idea of using NTFs for ENA imaging.

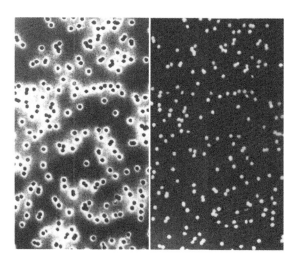

Fig. 7.7. Scanning electron microscope images of a nuclear track filter. Photographs credit: Alexander Mitrofanov.

After settling in California, I continued to search for alternative technologies for filtering EUV radiation out in ENA sensors and even published a short review on the subject.[29] In a couple of years, I found the so-called transmission gratings developed for a space X-ray observatory, *Advanced X-ray Astrophysics Facility*, or AXAF. NASA later renamed it the *Chandra X-ray Observatory*.

---

29. Gruntman, 1991b

7. Energetic Neutral Atoms Mike Gruntman

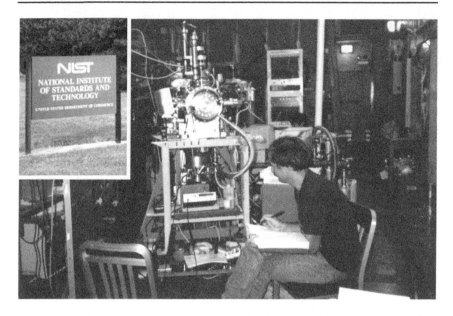

Fig. 7.8. Author at the *Synchrotron Ultraviolet Radiation Facility*, SURF II, at the National Institute of Standards and Technology, Gaithersburg, Md., in 1996. Photographs from collection of Mike Gruntman.

A free-standing modification of a transmission grating consisted of a set of parallel 100-nm-wide gold bars with geometric transparency of one-half. An extra large-mesh grid structurally supported the bars, which resulted in overall geometric transparency of about 30%. Mark Schattenburg and colleagues developed unique technologies for making such free-standing gratings at the *Massachusetts Institute of Technology*.

With NASA's support, I built an experimental setup to explore the transmission and polarization properties of the gratings at USC. The *National Institute of Standards and Technology*, NIST, in Gaithersburg, Md., also provided access to its electron synchrotron for tests (Fig. 7.8). The gratings demonstrated significant suppression of EUV and UV radiation.[30] ENA instruments would rely on such transmission gratings for filtering out ENA fluxes from the superior background of energetic photons on NASA's missions IMAGE (MENA) and TWINS launched afterward.[31]

The openings in transmission gratings were slits and not cylindrical holes. Consequently, they also exhibited strong polarization properties.

---

30. Gruntman, 1995a,b; 1997a,b
31. Pollock et al., 2000; McComas et al., 2009a

The gratings could thus uniquely serve either for producing polarized radiation in UV and EUV or for measuring the polarization of radiation in this spectral region.

The main newsletter of SPIE (The *International Society for Optical Engineering*), *OE Reports*, described highlights of its annual International Symposium on Optical Science, Engineering, and Instrumentation in 1995:

> In the X-Ray Spectrometers and Instrumentation session (Conf. 2515), chair Marilyn Bruner felt that the acme of her session was a discussion born of a paper given by Mike Gruntman (University of Southern California, Los Angeles, CA), on transmission grating filtering and polarization characteristics in EUV. According to Bruner, the search for a UV polarizer has been decades long and the need for such a device is finally satisfied. Using this small transmission grating at an entrance slit of a spectrograph allows for polarimetry, a technology that has not heretofore been available.[32]

While the development of ENA filters took place in California after I had left IKI, the idea and motivation grew directly from the original 1979 List and early work on neutral atom instrumentation.

ENA fluxes at 1 AU and in interstellar space

"Real" theoreticians did not pay attention to ENAs, but somebody had to establish possible ENA fluxes to guide instrument development. This pleasant "chore" provided occasional fun for me. The theoretical excursions were not easy, however, because of the perennial lack of time. Besides, my boss also frowned on such studies and tried to keep me focused exclusively on experimental work, preferably on atomic collisions.

Early estimates of the neutral fraction in the solar wind, the neutral solar wind, at the Earth orbit[33] provided indispensable guidance for developing instrumentation. Charge exchange of the solar wind protons on interstellar gas flowing into the solar system forms the neutral solar wind. Its fraction was small, 10^{-4}–10^{-5}, at 1 AU from the Sun. The predicted neutral atom flux also varied along Earth's orbit during the year, which depended on the direction of the interstellar wind velocity and temperature of the interstellar gas in the LISM. Therefore, measuring the neutral solar wind would probe the properties of the interstellar gas.

In addition, I realized at that time that the neutral fraction in the solar wind would become large, 10% or more, at the interstellar boundary of the

---

32. Cowan, 1995
33. Gruntman, 1980b

solar system. This NSW flux would thus cross the solar system frontier and enter and disturb the local interstellar medium (Fig. 7.5). There, these ENAs would charge exchange on interstellar plasma ions, imparting momentum and energy to the approaching interstellar wind. Therefore, the Sun "contaminates" the surrounding interstellar medium to distances much larger than had been thought.

Baranov's model of the global interaction of the solar system with the LISM was among the most advanced in the world at that time. Based on the best available observations, it considered the interstellar wind as supersonic. Consequently, it predicted a bow shock to be formed in front of the heliosphere (Fig. 7.5).

I showed that if one assumed the heliosphere geometry and parameters of the Baranov model, then ENAs of the neutral solar wind would slow down and heat the interstellar wind, making it subsonic. The flow pattern would consequently change substantially. Deflections of the initially straight streamlines of the interstellar wind flow approaching the bow shock, shown in Fig. 7.5, are due to this effect. Fundamentally, the supersonic interstellar plasma flow would thus sense the presence of an object (the heliosphere) ahead even without undergoing a shock transition. This was an unexpected and important new feature of the global heliospheric interaction.

I presented the findings at the department's science seminar, with Vladimir Baranov presiding. At the end of the talk, Ruderman asked whether my effect invalidated the initial assumption of the geometry of the heliosphere outlined by the Baranov model. I answered cheerfully in the affirmative.

The mathematically rigorous culture dominated the seminars. It originated from the Department of Mechanics and Mathematics of the Moscow State University, where Baranov lectured. Many staff scientists in his IKI group were the department's alumni. They usually reduced all problems to dimensionless form and applied advanced mathematical techniques. The participants frowned upon rare talks like mine based on estimates and physics arguments. There were no objections, however, to my results.

I described the effect of the neutral solar wind on the global heliospheric interactions in an article[34] in *Pis'ma v Astronomicheskii Zhurnal* in 1982. This journal was cover-to-cover translated to English as *Soviet Astronomical Letters*; it was a Soviet analog of *Astrophysical Journal Letters* in the United States. Afterward, Baranov liked to emphasize that he did not excommunicate me from the temple of department theoreticians despite my "disrespect" of his model.

---

34. Gruntman, 1982

In 1993 Baranov and Malama made a major step in heliosphere modeling by accounting for charge-exchange coupling between the interstellar neutral gas and plasma in the interstellar wind self-consistently. The computational capabilities had improved by that time, and Malama (Fig. 7.4) implemented highly efficient Monte Carlo techniques. They also, for the first time, added my neutral solar wind effect to the global interaction and confirmed its importance. Since that time, theoretical studies of the heliospheric global interactions routinely include the effect into modeling.

I originally assumed that the interstellar wind plasma instantaneously absorbed the energy and momentum of the charge-exchanged neutral solar wind atoms. An IBEX team suggested later a refinement, pointing to a possibility for newly formed protons in the interstellar plasma to preserve their pitch angle in the local interstellar magnetic field for some time.[35] This hypothesis could explain, at least in part and if valid, the still not completely understood ENA ribbon observed by the IBEX mission at the interstellar boundary of the solar system and shown on the cover of *Science* magazine (Fig. 3.8).

On a free range

Exploring ENAs was exciting and interesting, as the entire field was new, with practically no publications in journals. At the same time, it was also lonely and difficult to work without anybody around to discuss the science.

My boss, Leonas, carried the main burden of listening to my brainstorming ideas. This must have immensely annoyed him as they distracted him from atomic collisions and differential scattering. Sometimes, he helped me with advice. The most important thing, however, was that he had finally acquiesced to me doing what I wanted. He did not approve and frowned upon some of my pursuits. But I was roaming on a free range. Almost.

My diversions to unrelated applications of detectors and theoretical aspects of neutrals in space also led to periodic frictions between us. The above described neutral solar wind interaction with the interstellar wind could serve as a prime example. A few other projects fell into this category on the experimental side as well.

The work on MASTIF[36] for secondary ion mass spectrometry, described below (Chapter 8), resulted in really serious tensions. Leonas also did not initially support the experimentally challenging study of statistics of ENA-

---
35. Heerikhuisen et al., 2010
36. Gruntman, 1989

produced electron emission from ultrathin foils.[37] Another disapproved project was my unsuccessful attempt to introduce time-of-flight techniques to mass analysis in multichannel energy analyzers for passive corpuscular diagnostics of fusion plasmas.[38] (Such neutral particle analyzers usually combined ion analysis in electrostatic and magnetic fields.) I could not convince a group at LFTI in Leningrad that "owned" that area[39] to adopt the new approach. Some consolation came 10-20 years later when the *Joint European Torus*, JET, fusion plasma facility in Culham, England, implemented a similar technique.

I always respected my boss and kept him informed about my expanding activities. It became easier for Leonas to cut me loose on a free range rather than control me. It was simply not realistic to keep me on a tight leash in any case. Besides, my work had been bringing visibility to the laboratory. I am truly grateful that Leonas allowed me to work on energetic neutral atoms in space and all kinds of related and many unrelated problems, away from the core field of atomic collisions.

It is true that working alone on ENAs took a toll and sometimes looked like a road to nowhere. It was very possible that my interests in energetic neutral atoms would have died out with time because of lack of support, with the activities turning to other fields. Certain fortunate events, however, interfered in the year 1979, the same year of the compilation of my 1979 List, and changed the situation dramatically.

Help came from abroad.

---

37. Gruntman et al., 1989, 1990a
38. Gruntman, 1983
39. Petrov et al., 2021

8. First Steps Together

Foreign entanglements

It was the year 1979 when the first links extended to colleagues outside the Soviet Union who were interested in energetic neutral atoms in space. At that time there were very few of those in the world. One could count them on the fingers of one hand.

The Cold War was in full swing. The Communist Party of the Soviet Union tightly controlled the society and the people. Supreme leader Leonid Brezhnev (Fig. 8.1) and the Politburo ruled one-third of the world. The Soviet troops and soldiers of the satellite countries faced the Americans and their allies in the heart of Europe (Figs. 8.2, 8.3). During those dark years, IKI director Roald Sagdeev embarked on expanding foreign collaboration by flying Western science instruments on Soviet spacecraft.

In IKI's Department No. 18, personal attitudes of Leonas and Baranov facilitated the establishment of cooperation. Both laboratory heads were critical of the suffocating Soviet system with its inherent isolation from other countries in general and the science community in particular. Leonas welcomed my desire to connect with neutral-atom and detector colleagues in Europe and the United States, often in the areas outside the fields of his scientific interests. We took full advantage of the possibility to invite foreign scientists as guests of the USSR Academy of Sciences (Chapter 6), which also aligned with Sagdeev's promotion of international cooperation.

Among our very first neutral-atom contacts were Hans Fahr, a professor at the *University of Bonn* in West Germany, and Stanislaw Grzedzielski (Stanisław Grzędzielski), director of the *Space Research Center* (*Centrum Badań Kosmicznych*, CBK) of the Polish Academy of Sciences in Warsaw.

8. First Steps Together — Mike Gruntman

Fig. 8.1. Statue of Soviet supreme leader, 1964-1982, Leonid Brezhnev in an arts park museum in Moscow. Photograph (2018) courtesy of Mike Gruntman.

The world map is very different today. In October 1990, West Germany, or the Federal Republic of Germany (FRG), formed a unified Germany by absorbing East Germany, the communist German Democratic Republic (GDR).

Hans Fahr (Fig. 8.4) visited us in 1979. He became the first guest hosted by the group of Leonas in an area of space research and unrelated to atomic collisions. Hans and I established excellent scientific relations, lasting to this day, since we had met for the first time at the Moscow international airport Sheremetyevo. We became good friends.

Fahr published several important theoretical articles on the inflow of interstellar gas into the heliosphere in the late 1960s. In particular, he introduced the treatment of interstellar atoms as individual particles flying along hyperbolic orbits in the solar system. This approach provided important insight into their characteristics, essential for the direct detection of interstellar helium. Hans also considered the neutral solar wind.

At that time, Fahr worked closely with his older colleague at the University of Bonn, Peter Blum (Fig. 8.5). In the 1960s and 1970s, they published a series of "Blum and Fahr" and "Fahr and Blum" articles on various aspects of the solar system interaction with the surrounding interstellar medium.

Stanislaw, or Stan, Grzedzielski (Fig. 8.6) directed a small (by Soviet standards) institution in Warsaw. The Space Research Center, CBK in Polish, opened in 1977. From the first days, it had been expanding its space research projects and actively collaborating with IKI. Today, CBK em-

Fig. 8.2. Left: iconic U.S. Army Checkpoint Charlie in Berlin today. It separated the American sector of West Berlin from communist East Germany, GDR, during the Cold War. Right: preserved part of the Berlin Wall erected by GDR to keep its people from escaping the beloved socialist paradise by barbed wire, land mines, guard dogs, and bullets. Photographs (2011) courtesy of Mike Gruntman.

Fig. 8.3. Preserved warning sign near Checkpoint Charlie in Berlin. Photograph (2011) courtesy of Mike Gruntman.

Fig. 8.4. Hans Fahr, University of Bonn, in Bad-Honnef, Germany, in 2005 at the event dedicated to his mandatory retirement and transition to the emeritus status, *Emeritierung*. In addition to multiple scholarly articles in space physics and astrophysics, Fahr authored the book *With or Without the Big Bang. That's the Question* (*Mit oder ohne Urknall. Das ist hier die Frage*). Photograph from collection of Mike Gruntman.

Fig. 8.5. Left: Peter Blum (left) and Hans Fahr at the Space Gas Dynamics Conference in Moscow in October 1988. Right: Fahr (right), Blum, and Gruntman (left) 16 years later at a lunch celebrating Peter's 80th birthday in Bonn in June 2004. Photographs courtesy of Mike Gruntman.

ploys two hundred staff members. Importantly, the personal science interests of Grzedzielski included heliospheric neutral atoms. He would serve as an effective and consequential advocate for the entire field.

Our Polish colleagues lived and worked under substantially lighter political restrictions and constraints than we in Moscow. Consequently, Stan played an important "interface" role in maintaining communications with our Western colleagues. This turned out to be particularly valuable for our ENA work because of restrictions on my travel abroad. Stan and I joked, playing on words, about a neutral role (between the East and West) served by the neutrals (neutral atoms) studied by the Warsaw group.

Fig. 8.6. Stan Grzedzielski served as director of the Space Research Center of the Polish Academy of Sciences in Warsaw throughout the 1980s. Photograph (2005) from collection of Mike Gruntman.

The director of the leading Polish space research institution Grzedzielski had access to IKI director Sagdeev. This communication channel became especially important for our space effort because both Leonas and Baranov did not get along with Sagdeev. Stan mitigated many negative consequences of this state of internal affairs at IKI.

One sentence in a report

Sometime in the early spring of 1979, in the IKI library, I stumbled on a recent annual report on space research activities by the Federal Republic of Germany. Perusing this document, I saw a sentence that would have a major impact on me. The sentence simply said, in the passive voice, that a possibility of direct detection of neutral atoms in interplanetary space was being investigated. This one lone sentence, nothing more, provided a huge boost to my morale. Now I knew in my isolation that there was somebody else in the world interested in directly detecting individual neutral atoms. After all, I was not alone in this Universe!

8. First Steps Together

Soon in April of that year, Ian Axford gave a lecture at a science seminar at IKI. Axford was one of the directors of the *Max Planck Institute for Aeronomy* (*Max-Planck-Institut fuer Aeronomie*, MPAe) in Katlenburg-Lindau in West Germany. Located 40 km north of Goettingen in Lower Saxony, the institute became one of the most active research centers in space science. It had four science codirectors, who rotated to serve in the administrative capacity.

Fig. 8.7. Helmut Rosenbauer in the early 2000s at MPAe, Katlenburg-Lindau. He served as an institute director from 1977-2004. Photograph courtesy of Mike Gruntman.

After his talk, I approached Axford and asked about the detection of neutral atoms mentioned in the recent FRG report. Ian immediately explained that his colleague and MPAe codirector Helmut Rosenbauer (Fig. 8.7) had been working on this problem. Fortunately, theoretician Axford knew about such an esoteric and minor (at the time) effort in instrument development. Axford's interests concentrated on the heliosphere from the point of view of global plasma flows, magnetic fields, and especially energetic particles and cosmic rays. Interstellar neutral atoms existed only on the margins of his world.

I sent a letter to Rosenbauer and included, to justify mailing, a few of my publications. I wrote about my interest in directly detecting interstellar helium atoms and ENAs in general. (Parenthetically, all letters going abroad had to be approved first by the head of the laboratory or department at IKI followed by the institute's security and foreign-relations administrative staff. Therefore, the text in the letters had to be often oblique and a bit foggy.) I did not hear back from Germany for several months.

Suddenly, a lengthy reply letter from Rosenbauer arrived, two or three pages of single-spaced text. First, he apologized for the delay with the response because of the pressing schedules in many projects. Then, he described in detail his work on the direct detection of interstellar helium atoms. His approach relied on secondary ion emission, a technique shown in Fig. 7.6,d, right. An apology for the delay struck me. No institute director would have ever done it, even perfunctorily, in the USSR.

I quickly wrote Rosenbauer back, expressing, not very diplomatically, doubts about his detection concept but wishing success. The director of the Max Planck Institute must have been surprised to hear such words from a young scientist about whom he had never heard just a few months ago.

A month later, the door to my laboratory-cum-office at IKI suddenly opened, and Rosenbauer walked in, accompanied by a terrified scientist from another department of the institute. This first meeting began our life-long collaboration and friendship.

The colleague who brought Rosenbauer to see me felt very uneasy, as this encounter was not in the program. A visit to IKI by foreign scientists required writing internal memos and producing various documents that had to be approved well in advance by the foreign and regime departments of IKI, then by institute deputy directors, and finally by the Directorate of Foreign Affairs of the Academy of Sciences. The paperwork described in minute detail where and when the visitors would go, who they would see and meet, and the topics to be discussed. The guests had to always be accompanied by designated scientists identified by name who would be responsible for taking them from one approved meeting or event to another.

IKI director Roald Sagdeev energetically expanded foreign cooperation by flying Western scientific instruments on Soviet spacecraft. Such projects required frequent visits and meetings between participating scientists and engineers. Consequently, the number of guests had been rapidly growing. The ossified control system simply could not effectively oversee the stream of visitors. Besides, the socialist top-down bureaucracy was inherently ineffective and dysfunctional, which added to the system's stress and strain.

Sagdeev certainly deserves major credit for promoting and growing collaboration in a suffocating totalitarian environment. He opened a window to the world—perhaps not a large one, but a very useful and consequential pane.

Meeting Polish and West German colleagues was a stroke of truly great luck for me. It prevented my ENA work from fading away due to a lack of institutional interest and support. By the mid-1980s, we had established contacts with several scientists interested in the direct detection of neutral atoms. Among those was also an American, K. C. (Johnny) Hsieh (Fig. 8.8), a physics professor at the University of Arizona in Tucson.

8. First Steps Together — Mike Gruntman

Fig. 8.8. Johnny Hsieh (left) and Stan Grzedzielski (right) at the Fall Meeting of the American Geophysical Union in San Francisco in December 1990. Photograph courtesy of Mike Gruntman.

Johnny Hsieh was among the very few pioneers, such as Helmut Rosenbauer in Lindau and us in Moscow, who accepted the experimental challenge of developing dedicated ENA instrumentation in the late 1970s.[1] Unfortunately, his role and contributions to advancing the field in those early days are rarely given the credit they deserve.

Already in 1980, Hsieh submitted a proposal to NASA to conduct in situ measurements of neutral atoms in the geospace and the heliosphere.[2] Special professional and personal relations developed between ENA pioneers based on common interests in this emerging field. Importantly, Stanislaw Grzedzielski and his group in Warsaw also became an integral part of this effort in the very early 1980s. With time, ENAs would evolve into a competitive field dominated by large, powerful institutions. As one very keen observer of people and events put it, "[T]he men who go first are rarely popular with those who wait for the wind to blow."[3]

Hsieh and I first met when he visited IKI with Charles Curtis (Fig. 8.9) from his group in the early 1980s. They participated in an experiment on the Soviet Vega mission to comet Halley. A tall man, Curtis, showed up in my laboratory room in huge, as they appeared to me, hiking boots. Next to Johnny, he looked like a giant. Hsieh and Curtis would actively continue studies of energetic neutral atoms for twenty years.

Johnny and I quickly established excellent working relations. Our friendship lasts to this day. A mere four months after I had joined USC in March 1990, I drove to Tucson to work on a joint proposal with him

1. Gruntman, 1997a, p. 3626
2. Hsieh, 1980
3. Jones, 1978, p. 265

Fig. 8.9. Charles Curtis in Arizona in July 1990. Photograph courtesy of Mike Gruntman.

(Fig. 8.10). This trip was memorable for another reason as well. My first immigrant car bought for eleven hundred dollars did not have air conditioning, a noticeable feature in Arizona in July.

By the early 1980s, my lonely days had come to an end. It became possible to discuss the ideas and exchange publications with several foreign colleagues. Communications in the pre-Internet era also presented difficulties. Tight security and ideological control on the communist side of the Iron Curtain did not help either. The neutral atom colleagues were far away in foreign lands, but the collaboration took off.

Fig. 8.10. The author of this book with Johnny Hsieh (left) on the campus of the University of Arizona in Tucson on a hot day in July 1990. Photograph from collection of Mike Gruntman.

8. First Steps Together   Mike Gruntman

Professional growth

Rosenbauer visited IKI frequently, overseeing a number of his joint space experiments. He worked primarily with the group of Konstantin Gringauz. IKI usually assigned junior or middle-level scientists to accompany a foreign guest. They were afraid to say no to Helmut when asked to take him to see me in my laboratory, which was outside the approved joint work program.

Fig. 8.11. Helmut Rosenbauer with an early engineering mockup of a comet lander, Philae, of the Rosetta mission in the early 2000s at MPAe. Photograph courtesy of Mike Gruntman.

MPAe supported internal travel inside FRG and paid per diem to Soviet visitors. An MPAe director, Rosenbauer (Fig. 8.11), could influence who among Soviet partners should visit his institute for work on their joint experiments and for how long. For many Soviet specialists, such trips, especially to wealthy West Germany, presented the most valued perks, rewards, and incentives.

Per diem for a week-long trip could bring highly desired clothing, consumer electronics, and other goods. At that time, a pair of Western jeans cost $20 in Europe and fetched an average one-month salary on the black market in the Soviet Union. In addition, one viewed oneself as part of an elite, anointed and trusted for travel to Western countries. The desire and struggle to get on foreign trips even in a capacity for which a scientist or engineer was overqualified were demeaning and humiliating.

Many opportunities for such trips occurred on joint projects to deliver parts of space instruments or flight electronics, usually in well-built sturdy boxes, to or from MPAe for integration and testing. I heard privately more than one friendly German colleague casually refer, derisively, to some traveling IKI specialists as *Gepaecktraeger*. The word meant a porter, a luggage carrier at a railroad station.

With time, Rosenbauer became familiar with the IKI building and began coming to see me without an accompanying escort scientist and without telling his hosts. Such visits violated the sacred rules. Helmut clearly liked breaking them.

Rosenbauer was an excellent physicist. We enjoyed our long meetings every two or three months. He became a trusted close colleague with whom I could regularly brainstorm and openly discuss in detail my many ideas and experimental work. His feedback, genuine interest, and support

Fig. 8.12. Institute director Rosenbauer "riding into the sunset" dressed as an Indian chief at a staged ceremony organized by MPAe to mark his retirement in summer 2004. Manfred Witte (in a white shirt) stands near the column on the left. Photograph from collection of Mike Gruntman.

8. First Steps Together   Mike Gruntman

were very helpful. Nobody else at IKI or elsewhere played such a role in my science life and professional growth in the 1980s. Many years later, Rosenbauer told me that the benefits of our exchanges were mutual. It was truly flattering. We became very good friends and remained so until his last days (Fig. 8.12).

Interstellar helium experiment

The instrument of Rosenbauer to directly detect individual atoms of interstellar helium relied on secondary emission of positive ions from a specially prepared and periodically refreshed onboard lithium fluoride (LiF) surface (Fig. 7.6,d, right). An impact of a 100-eV helium atom knocked out, with some probability, a Li^+ ion that was then detected by a miniature channel electron multiplier. Importantly, background ultraviolet photons of the dominating hydrogen Lyman-alpha line at the wavelength of 121.6 nm penetrated through the LiF layer, which resulted in low photoelectron emission from the sensitive surface, reducing overall sensor noise.

On September 10, 1972, Rosenbauer submitted a comprehensive proposal "Direct Measurement of the Parameters of the Interstellar Gas" (Fig. 8.13). He outlined an experiment on a heliocentric space probe "to determine the density and temperature of the interstellar gas and its velocity relative to the solar system."[4]

Rosenbauer was then on staff of the *Max Planck Institute for Extraterrestrial Physics* (*Max-Planck-Institut fuer extraterrestrische Physik*, MPE), which was a sub-institute (Teilinstitut) of the *Max Planck Institute of Physics and Astrophysics* (*Max-Planck-Institut fuer Physik und Astrophysik*) in Garching, near Munich.

The proposal listed two coinvestigators, Hans Fahr from the University of Bonn and William (Bill) Feldman from the *Los Alamos National Laboratory* in New Mexico. Many years later, Bill told me, very modestly and likely understating, that he was just a visitor in Garching and participated in discussions of the concept.

At the time of the proposal submission by Rosenbauer, I was a second-year student, a sophomore, in a faraway land behind the Iron Curtain and obviously had no clue about interstellar gas, trajectories of helium atoms, and detectors. Remarkably, in a mere three years, I would "discover" this challenge independently and start working on it. After my master's thesis (Fig. 3.5), the circumstances shifted the effort to detection of hydrogen ENAs with energies of a few hundred electronvolts and higher.

---

4. Rosenbauer, 1972, p. 3

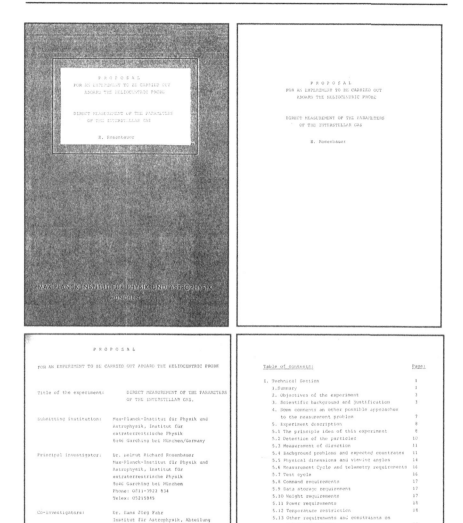

Fig. 8.13. First four pages of the proposal "Direct Measurement of the Parameters of the Interstellar Gas" by Principal Investigator H. Rosenbauer, dated September 10, 1972.

8. First Steps Together   Mike Gruntman

Rosenbauer's experiment, named *GAS*, was originally scheduled for launch on the Ulysses spacecraft in 1986. The tragedy of the Space Shuttle *Challenger* delayed its launch. Ulysses finally left the Earth on the Space Shuttle *Discovery* on October 6, 1990. (Note that the Soviet-German-Polish neutral-atom experiment, described in Chapter 9, was also named GAS. To avoid confusion, I italicize the name *GAS* of the sensor on Ulysses.)

By October 1990, I had firmly settled in Los Angeles. Helmut came to NASA's *Jet Propulsion Laboratory*, JPL, in nearby Pasadena for turning on the instrument. Initially, they had difficulties opening the cover of the *GAS* sensor. It was a very stressful time for him. Finally, after several days and trying various experimental "tricks," the system worked as designed and the instrument became operational.

Fig. 8.14. Helmut Rosenbauer at a campfire in Sequoia National Forest in California, celebrating successful activation of the interstellar helium atom instrument *GAS* on Ulysses in late October 1990. Photo courtesy of Mike Gruntman.

Helmut called me at USC with this great news and proposed to celebrate the success. As an avid hiker, he thought about the mountains. So, in a day or two, we headed to Sequoia National Forest, about 150 miles away. On the way, we stopped in Bakersfield to buy steaks, other food, and adult beverages. We camped overnight under the stars, roasted steaks on the campfire, and celebrated the success of the *GAS* experiment (Fig. 8.14). It took an effort of many people and 18 years from Helmut's 1972 proposal to the Ulysses launch. This pioneering experiment to directly detect interstellar helium atoms would become a resounding success.

MASTIF for interstellar helium and SIMS

Reliable detection of weak neutral-atom "signal" counts and suppression of high background (noise) caused by ultraviolet photons, cosmic rays, and intrinsic radioactivity of sensor materials were and are major requirements for all ENA instruments. Time-of-flight techniques (as in Fig. 7.6,c) essentially relied on double (sometimes triple) coincidences which effectively suppressed noise.

In 1981, after familiarizing myself with the details of the interstellar helium sensor *GAS*, I proposed to Rosenbauer to add coincidences in an unusual way[5] for suppressing the noise. Then, in 1982 it became clear to me that such TOF-based coincidences could also provide position sensitivity. The latter feature, in turn, would have also enabled concurrent measurements of directional intensities of interstellar helium atoms in a wide field of view (FOV) rather than scanning the sky with a narrow, "pencil" FOV single-pixel sensor.

My unconventional time-of-flight idea had never been tried in laboratory experiments. NASA and European Space Agency (ESA) fixed the original launch date of Ulysses for 1986. Consequently, the instrument delivery schedules precluded making changes in *GAS*. I also felt that Helmut did not believe in my new concept to improve the sensor. I did not press it, being busy with other projects. Finally, in 1986, I squeezed out the necessary time and designed and built a set of elegant detectors, demonstrating the idea in the laboratory.

The new technique also enabled an unusual approach to secondary ion mass spectrometry, or SIMS, of surfaces. Physics laboratories and industry widely used SIMS in various applications. My detectors demonstrated for the first time multichannel mass identification of secondary ions using continuous low-intensity probing beams. The latter opened a way for studying

---

5. Gruntman and Morozov, 1981, pp. 39, 40

8. First Steps Together

fragile and extremely thin objects such as, for example, ultrathin carbon foils, sensitive to damage by intense beams and high doses common for conventional SIMS. In addition, one could also characterize and monitor, for the first time, the contamination of sensitive surfaces of microchannel plates under real operational conditions in a vacuum.

After some pleasant struggles, I coined the acronym MASTIF which stood for mass analysis of secondaries by time-of-flight technique. I could not find a proper word to add the second letter "F." Even with one "F," the acronym was close enough to a large sympathetic dog.

It took me two more years to put the results into writing and submit an article to the American journal *Review of Scientific Instruments* in June 1988. I sent the manuscript "illegally," without the required permissions of the institute and the Academy of Sciences. Even Leonas did not know about it. He strongly objected that I spent time on MASTIF, with the resulting tensions between us.

I simply ignored the rules requiring the permissions. The bureaucratic system, always dysfunctional to some degree, began producing multiplying glitches under the currents and stresses of perestroika and glasnost. Many foreign colleagues who visited IKI knew me personally. Therefore, it was not difficult to find somebody willing to take the revised manuscripts with my replies to the editors for mailing them on return to their home institutions. Such arrangements caused some non-critical delays in communicating with the journal.

Monitoring the IKI incoming foreign mail also became less strict. In the 1970s and 1980s, a designated IKI staff member with knowledge of a foreign language opened all incoming letters from abroad and recorded in a logbook the basic information with the names of the letter author and recipient and the letter content. Humans are humans everywhere; they are also often lazy and seek favors. Even during the dark hopeless Soviet period of Brezhnev and Andropov, the staff member often gave letters arriving to my colleague Kalinin and me unopened.

She was a quiet middle-aged woman in an office nearby. A couple of times each year, Kalinin gave her a present, a small flask of alcohol for "personal purposes." So we were "friends." Corruptive laziness and lack of accountability inevitably permeate all bureaucracies and socialist systems in particular. Consequently, nobody noticed my incoming mail about an unauthorized article that appeared in *Review of Scientific Instruments* in October 1989.[6]

---

6. Gruntman, 1989

With an offprint of the just-published article in hand, I walked to the editorial office of the Russian cover-to-cover translation of this journal, *Pribory dlya Nauchnykh Issledovanii* (see also Chapter 5). I made an offer to the overworked and underpaid chief editor to translate my own article into Russian. Many scientists in Moscow moonlighted by translating articles for Russian editions of foreign journals to complement their income. For me, this was the time of final preparations for disappearance from the country in a few months. I needed money for travel.

The alert editor immediately asked me about the Russian version of the manuscript. I replied that I had written the article in English. With visible concern on his face, he pointed out that the Academy of Sciences required a review of the Russian original of the text for permitting a publication abroad. I simply told him that I had not asked for permission.

An awkward silence followed. He then hired me on the spot to translate the article, which I quickly did and was promptly paid for. The accelerating decay of the Soviet system and practical operational pressures outweighed his other considerations. The translated article soon appeared in the Russian edition (Fig. 8.15). The payment covered one-half of the price of the airline ticket to leave Moscow forever in a couple of months.

My article in *Review of Scientific Instruments* seemed to be the first-ever published in this journal by someone from the Soviet Union. There might

Fig. 8.15. Cover of the October 1989 issue of the translated Russian edition of the journal *Review of Scientific Instrument* with the MASTIF article. Publishing of this cover-to-cover translation ceased in July 1991 because of mounting financial losses. The value of translations also diminished in general, with knowledge of some English becoming almost obligatory in hard sciences and engineering.

have been an article or two with Soviet coauthors. But nobody published there directly.

Ten years later I felt immensely proud when the *American Institute of Physics*, AIP, confirmed me as a member of the editorial board of *Review of Scientific Instruments*.

Science is essentially an international endeavor. Neutral atoms in space were no exception. This new field would emerge with scientists from several countries leading the way.

9. ENA Experiment That Never Flew

Ultrathin foils for TOF instruments

As our focus turned to hydrogen ENAs with energies of a few hundred electronvolts and higher, the time-of-flight approach (Fig. 7.6,c) looked promising. Already by 1978, I had developed the basic concept of compact TOF sensors for measuring ENAs. It relied on microchannel plate detectors and freestanding ultrathin carbon foils, about 15-20 atomic layers thick. Nuclear physics used the foils in experiments with ion energies 20 keV and higher. Physics did not prohibit lowering the energy of atomic particles that could be analyzed by this technique. Therefore, its demonstration for ENA energies expected in space became a main task in the 1979 List (Chapter 7).

Making ultrathin carbon foils was and is an art as much as a science. Very few places and people in the world produce them. The fabrication process begins with the deposition of the foil material on a glass flat or microscope slide covered by a soluble substrate. For example, the evaporation of carbon from an arc *in vacuo* serves as a source of carbon. After dissolving the substrate in distilled water, the floating foil is picked up ("fished out") on a high-transparency supporting metal mesh. The foil attaches to the mesh with remarkable strength, capable of withstanding the adverse dynamic environments of a rocket launch. One often purchases a glass flat with a deposited foil and then carefully transfers it on a supporting mesh in a simple laboratory setup (Fig. 9.1).

Yuri Gott from the Kurchatov Institute of Atomic Energy gave me the first thin foils for my experiments in 1977. He would later serve as an official reviewer, an opponent, of my Ph.D. thesis. Vitaly Liechtenstein (Fig. 9.2) from a different division of IAE provided foils after 1980.

9. ENA Experiment That Never Flew — Mike Gruntman

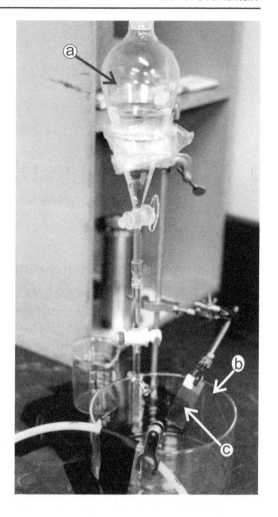

Fig. 9.1. Simple laboratory setup for transferring a deposited ultrathin foil from a glass flat to a supporting mesh assembled by the author of this book in the early 1990s.
(a) – Distilled water reservoir. (b) – Chemical glass beaker with a slowly rising or lowering water level.
(c) – Glass flat with a deposited foil.

As the substrate dissolves in the slowly rising water, the foil gradually peels off and finally floats free on the water surface. Then, one carefully raises the mesh in the water underneath the foil (or slowly drains water through a white tube seen at the bottom left of the photograph) and captures it. The foil sticks to the mesh.

Photograph courtesy of Mike Gruntman.

Vitaly was an excellent scientist and became a good friend. With the Soviet Union no more, he came as a visiting scientist to the U.S. Department of Energy's *Brookhaven National Laboratory*, or BNL, in the mid-1990s. I spent long hours on the telephone with him, helping with advice on establishing his programs and new life in the United States. Had he stayed at BNL, Liechtenstein could have become an important source of thin foils for physics groups in the country. Vitaly finally went back to Moscow because of his family situation.

Laboratory experiments with my new TOF detectors in 1978-1981 demonstrated the detection of weak fluxes of hydrogen-atom ENAs with energies

Fig. 9.2. Left: Vitaly Liechtenstein (right) surprised the author of this book by meeting and greeting him at the Moscow international airport Sheremetyevo when the latter came to Russia for the first time as a foreigner in the mid-1990s. The Soviet Union was no more by that time. Right: Liechtenstein in his office in the Brookhaven National Laboratory. Photographs from collection of Mike Gruntman.

down to 600 eV.[1] (My experimental setup could not produce monoenergetic beams of particles with energies below 600 eV.) The sensors also provided velocity analysis of ENAs and showed respectable detection efficiency of neutral atoms. In addition, the time-of-flight techniques in the nanosecond range inherently suppressed the noise count rate from randomly arriving background ultraviolet photons.

The first of the main tasks from my 1979 List (Chapter 7), lowering the energy limit of particles detected by thin-foil-based TOF instruments by one and a half orders of magnitude, had thus been achieved.

On the way to Phobos, sort of

The time had come to attempt detection of ENAs in space. The neutral solar wind represented a primary target and the experiment concept had been ready by 1981-1982.[2] This was also a decisive moment for my boss, Vladas Leonas, to get actively involved administratively. So far, I could

---

1. Gruntman and Morozov, 1981, 1982
2. Gruntman and Leonas, 1983

9. ENA Experiment That Never Flew   Mike Gruntman

quietly work in the laboratory alone, but getting to space called for political muscles. Fortunately, Stanislaw Grzedzielski and his Space Research Center had also become indispensable partners in advancing the neutral solar wind experiment by that time.[3]

A few trips of "our" academician Petrov to IKI director Sagdeev led to adding the NSW instrument to space flight plans. The process did not include any detailed proposal or review of the concept. One feudal lord talked to another and struck a deal. The process was beyond my pay grade.

Initially, IKI included our experiment in a mission to Mars's moon Phobos, planned for launch in 1986. The mission offered an excellent platform for NSW measurements during the cruise phase on the way to Mars. I wanted, however, a second similar sensor to look away from the Sun to detect heliospheric ENAs from the interstellar boundary of the solar system. IKI soon authorized its addition to the instrument. We hoped that it could also measure energetic atoms from the magnetospheres of Jupiter and Saturn, an idea advanced by the Warsaw group.

Later, Rosenbauer's interstellar neutral helium sensor, similar to the one prepared for the Ulysses mission, would be included in the instrument package as well. This comprehensive joint Soviet-German-Polish experiment also acquired, at some point, the name GAS. In hindsight, the name of our experiment should have been chosen differently to avoid confusion with Rosenbauer's instrument on Ulysses. As Mike Lampton quipped during his visit to IKI in the spring of 1983, "Astronomical observations by the GAS experiment. We should call the field GAStronomy!"

Available mass and power for science experiments represented the most precious and contested resources on all space missions. As the instruments "fattened" and "grew" in the required mass and power during development, competition for limited resources led to periodic trimming of science payloads on the spacecraft. IKI removed the GAS experiment from the Phobos mission during the first such science reduction in 1984. We were fortunate to exit early. I later saw the pain and anguish of scientists and engineers who had been working on experiments for years to see them kicked out closer to the launch.

Heavy *Proton* space launchers sent two spacecraft to Mars on July 7 and 12, 1988. The first space vehicle, *Phobos 1*, failed during the mission cruise phase. The second, *Phobos 2*, was lost orbiting Mars. A commemorative stamp marked this space mission (Fig. 9.3).

The notion of principal investigators, or PI, did not exist in the 1980s.

---

3. Gruntman et al., 1990c

Fig. 9.3. Postal stamp commemorating the Soviet Phobos space mission launched in 1988. From collection of Mike Gruntman.

A so-called "science lead" and "instrument lead" of an experiment shared PI responsibilities. Only staff members in senior positions such as heads of laboratories or independent units served as science leads at IKI. Correspondingly, the institute first designated our laboratory head Leonas the science lead of the GAS experiment. At this time, development of the VEGA mission to comet Halley proceeded as well, with growing frictions between my boss and IKI director Sagdeev. Leonas oversaw the work of four staff members of the department on the dust particle impact analyzer PUMA (Fig. 2.5) for that mission. One day, Sagdeev removed him from the project. Baranov later wrote that the reason for the dismissal of Leonas "after two years of selfless and high-quality work" was his "too independent behavior" and that "these events turned into a personal tragedy [for Leonas]."[4]

These developments also directly impacted the inclusion of the neutral atom experiment in the Phobos mission. Baranov described that

> Sagdeev did not want to appoint Leonas as the science lead of the [GAS] experiment. He declared that he would put the instrument on the Phobos spacecraft only if I [Baranov] play that role. In order to save the experiment, Leonas asked me to agree to the arrangement.[5]

As the work on the GAS experiment progressed, our Polish and German colleagues sent a formal top-level proposal to the USSR Academy of Sciences to conduct this joint experiment. The Academy then forwarded the proposal to IKI for review. Baranov explained that

> at Sagdeev's request, I [Baranov] wrote a positive review of the proposal, and joint work of three countries on this unique experiment, related to measurements of neutral particles in the solar wind, began.[6]

In reality, the joint work had been going for a long time. The proposal served to formalize the existing arrangements and informal commitments in bureaucratic systems. In addition to IKI instruments, Rosenbauer was

---

4. Baranov, 2009, pp. 277, 278
5. Baranov, 2009, p. 296
6. Baranov, 2009, p. 297

9. ENA Experiment That Never Flew — Mike Gruntman

interested in measuring interstellar helium and helped with the baffle for the NSW sensor, Grzedzielski and Fahr supported theoretical aspects of the work, and engineers in the Space Research Center in Warsaw designed an electronic control unit. The formal proposal to the Academy allowed our foreign partners to "legitimize" their participation and secure national funding for their effort. Director Sagdeev played an enabling role in the plan, shrewdly orchestrating a review of the proposal by its instigators.

The entire arrangement at IKI defied logic. Leonas sincerely despised scientists, turned into political operators, who populated many important managing positions in the institute and beyond. As everywhere in the country, membership in the Communist Party served as an important qualification for filling these positions, often more consequential than technical or managerial merit. Theoretician Baranov shared the attitude of Leonas toward these operators to the full extent and was also often naive about administrative workings. Developing, building and flying instruments in space required coordinated efforts of many units. This could only be achieved by working with the existing people and powers in place, whoever and whatever they were.

Leonas and Baranov showed uncommon integrity and could sometimes take a stand on moral issues. It was commendable and highly admirable in the oppressive Soviet environment, but this did not lead to having things done. A space experiment called for numerous practical steps, small and large, to be accomplished on schedule. It also limited endless science discussions and relied on perhaps not perfect but "good enough" engineering decisions. And everything had to be done within the existing system, supported by adequate resources, and required working with the people in place.

IKI appointed me the "instrument lead" (*vedushchii po priboru* in Russian) for the GAS experiment. This was a common position for IKI scientists responsible for the design, fabrication, and testing of the instrument and the operation and data analysis of the experiment. My role encompassed much of the practical work except that I could not talk to the decision-makers, such as the institute director and his deputies, or sign any papers, which was reserved for the science lead.

Baranov became the formal science lead, per Sagdeev's requirement, and interface to the IKI directors. Leonas, in reality, continued to serve in this role informally and advised Baranov on the steps to take. Everything else, except the interaction with the administration, fell on my shoulders but without any administrative authority. Consequently, I had to bring every issue and question to Leonas first, and then it went to Baranov. Very often perfunctory approvals followed, but not always. It did not help either that Leonas and Baranov did not want to spend their time on uninteresting but essential administrative tasks dealing with people that they did not personally respect.

Soon after the first purge from the Phobos spacecraft in 1984, IKI offered us a ride on another new mission under development, *Relikt-2*. Initially planned for launch in 1987, the spacecraft had to reach the Sun-Earth Lagrange point L2 for studying the microwave background radiation. The orbit in this region 1.5 million kilometers away from the Earth in the antisolar direction offered unique advantages for ENA measurements. Our work continued.

GAS Experiment in the "heroic" ENA era

The GAS experiment now formally included three separate sensors, GAS-1, GAS-2, and GAS-3, and a common electronic control unit GAS-E.[7] It became a joint endeavor by IKI, the Max Planck Institute for Aeronomy, and the Polish Space Research Center.

Rosenbauer would provide the GAS-1 sensor, a copy of his instrument on Ulysses, to measure interstellar helium fluxes (of the type shown in Fig. 7.6,d, right). The time-of-flight instrument GAS-2, with triple coincidences (of the type Fig. 7.6,c), would measure the neutral solar wind (Fig. 9.4). It included a high-quality narrow field-of-view baffle allowing it to point only a few degrees away from the Sun. The GAS-3 sensor pointed in the antisolar direction to measure heliospheric ENAs from the interstellar boundary of the solar system. It was similar to GAS-2 but with a simple baffle and a wider field of view. In addition, it was hoped that GAS-3 could detect energetic neutral atoms emitted by the magnetospheres of Jupiter and Saturn. Both GAS-2 and GAS-3 included optional UV suppressing filters (Fig. 7.6,e), still to be developed.

Not only was Rosenbauer committed to providing the GAS-1 sensor by MPAe, but he also volunteered to design and build a baffle for GAS-2. He enjoyed this latter challenge and continuously worked on it in spare moments for several years. IKI contracted SNIIP's Khazanov and Gorn to build flight units GAS-2 and GAS-3 per our design. IAE's Liechtenstein supplied ultrathin foils for the instruments.

Grzedzielski's Space Research Center in Warsaw built the control unit GAS-E. CBK's Marek Hlond (Hłond) led the effort on that task. GAS-E served as the power and telemetry interface between all sensors and the

---

7. Gruntman et al., 1990c

9. ENA Experiment That Never Flew — Mike Gruntman

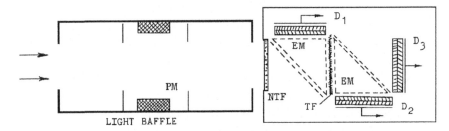

Fig. 9.4. Schematic of the first-generation direct-exposure thin-foil-based time-of-flight ENA instrument for the neutral solar wind (with a high-quality narrow field of view baffle) and heliospheric ENAs from the interstellar boundary of the solar system (with a simple baffle). PM – permanent magnet; EM – electrostatic mirror; TF – ultrathin foil; NTF – optional nuclear-track-type filter; D_1, D_2, D_3 – microchannel plate detectors. Schematic after Gruntman et al., 1990c, and Gruntman, 1997a.

spacecraft bus. Such an arrangement made it easier to interact with the security-conscious space industry that manufactured the spacecraft. Consequently, this part of the Soviet defense establishment worked primarily with IKI and engineers from Poland, eliminating contacts with the West German participants.

Directors of their respective institutions Rosenbauer and Grzedzielski were responsible for the German and Polish parts of the GAS experiment. This strengthened our position at IKI internally because they interacted directly with Sagdeev. Consequently, many burning questions and issues were resolved at IKI not by the science lead of the GAS experiment but by Helmut and Stan bringing them up to the IKI director.

Our cooperation on neutral atoms with the Polish colleagues ramped up in 1982-1983, especially after the space experiment concept emerged and IKI approved it as a flight project. Grzedzielski had been actively lobbying Sagdeev to let me periodically visit his Space Research Center in Warsaw, which was indispensable for the experiment and for advancing our neutral atom enterprise in general. Rosenbauer also strongly supported this. Finally, the officialdom relented and IKI and the Academy of Sciences allowed me to travel to Poland in support of the joint experiment. It was a fraternal country in the socialist camp from where no one could escape anywhere anyway. West Germany and other free world countries remained closed, verboten.

Helmut enjoyed annoying his Soviet counterparts. One day, in 1982 or 1983, he walked into my laboratory room grinning and in a great mood. He told me that he and another MPAe director, Axford, had a meeting with Sagdeev and several IKI laboratory heads and group leaders on joint experiments of the two institutes. Somehow, neutral atoms came into the discussion. Rosenbauer praised my work and argued that I should build flight instruments for their detection.

Sagdeev apparently pointed out that I was a laboratory physicist without flight instrument experience. "Not a problem," replied Rosenbauer, "send him to our Max Planck Institute in Katlenburg-Lindau. We will support him in MPAe for a year or two and he will learn and build the instruments there in West Germany." Axford also supported this suggestion. They both knew that the government would unlikely allow me to go.

"You should have seen the faces of IKI's group leaders," continued Helmut. "Many of them fight for an opportunity for a week-long trip to Germany. Now two Max Plank Institute directors offered to Sagdeev to send you there for a couple of years, with everybody tacitly understanding that you may never come back." Rosenbauer grinned as he concluded, "You certainly got a few more new jealous enemies today at IKI, but you probably do not mind."

Well, as Germans would say, "*Mehr Feinde, mehr Ehre*" (more enemies, more honor).

Our instrument never made it to space. Sensors GAS-2, GAS-3, and control unit GAS-E had been built by the early 1990s. MPAe could have provided the GAS-1 sensor whenever needed. The space mission Relikt-2 (Fig. 9.5), finally rescheduled for launch in 1993, never took off the launch pad as the entire country disintegrated and the space program nose-dived.

The GAS experiment was the first comprehensive dedicated space experiment of the "heroic" ENA era with the first-generation neutral-atom instruments to detect the neutral solar wind and heliospheric ENAs from the interstellar boundary of the solar system. I was glad that we briefly described this work at the First COSPAR[8] Colloquium "Physics of the Outer Heliosphere" in 1989 (Chapter 10). If it were not for this talk and associated publication[9] in the colloquium proceedings, the experiment would have been forgotten by now.

It is interesting that nobody has ever definitely measured the neutral solar wind fluxes and characterized their properties to this day. An instrument on the IMAGE mission observed enhancements in a response of a wide field-of-view sensor that might have been caused by ENAs in the

---

8. The *Committee on Space Research* (COSPAR) of the *International Council of Scientific Unions* is the major international umbrella organization of institutions and scientists engaged in space science research.
9. Gruntman et al., 1990c

Fig. 9.5. A mockup of the microwave instrument panel, dated 1993, of the Relikt-2 spacecraft in the IKI museum. In the second half of the 1980s, the institute planned to install sensors of the GAS instrument on this spacecraft. Photograph (2018) courtesy of Mike Gruntman.

solar wind. It could not unambiguously attribute the signal and determine the NSW properties in detail.[10]

Launched in late 2008, Interstellar Boundary Explorer did detect heliospheric ENAs from the interstellar boundary of the solar system and probed its properties.[11] NASA specifically selected and launched this Earth-orbiting mission to discover and remotely map the solar system frontier. Our early work at IKI began this pursuit in the 1970s and contributed to the development of the concept of such a mission and ENA instrumentation.

The road to IBEX was long. Heliosphere imaging in ENA fluxes was first included as part of a proposal for a larger *Interstellar Pathfinder* space mission that emerged from an initiative[12] started in 1998 (Chapter 10). The first proposal combined ENA imaging of the interstellar boundary with detailed measurements of the isotopic and elemental composition of the pickup ions in the solar wind. A team led by Principal Investigator George Gloeckler twice proposed *Interstellar Pathfinder* as Medium Explorers[13] in 1998 and 2001 (Fig. 9.6), without success. Then in 2003 NASA selected a Small Explorer (SMEX) mission, Interstellar Boundary Explorer, Principal Investigator Dave McComas, which focused on detecting heliospheric ENAs.

10. Moore et al., 2001; Collier et al., 2001
11. Funsten et al., 2009; McComas et al., 2009b
12. Moebius et al., 1998; Gruntman, 2015b
13. Moebius et al., 1998; McComas et al., 2003

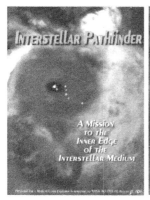 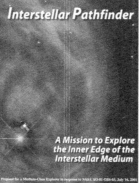 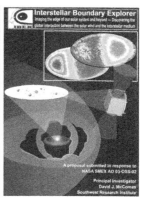

Fig. 9.6. Space mission proposals leading to the Interstellar Boundary Explorer, IBEX, launched in 2008. The mission would successfully map the interstellar boundary of the solar system (Fig. 3.8). Left: *Interstellar Pathfinder. A Mission to the Inner Edge of the Interstellar Medium* (Principal Investigator George Gloeckler), 1998. Middle: *Interstellar Pathfinder. A Mission to Explore the Inner Edge of the Interstellar Medium* (PI George Gloeckler), 2001. Right: *Interstellar Boundary Explorer. Imaging the Edge of our Solar System and Beyond,* (PI David McComas), 2003. From collection of Mike Gruntman.

The debrief of the successful IBEX proposal by the NASA program officers specifically mentioned that "the boundary between the solar wind and the local interstellar medium was one of the last unexplored regions of the heliosphere," that "[i]n the scientific merit, the proposal received very, very high marks" and there was "high probability of major ground-breaking science from this mission." They specifically pointed out that the proposal risk was "mitigated ... by the results of detailed modeling presented in the proposal and in the scientific publication in JGR."[14] This was the reference to our article in the *Journal of Geophysical Research* published in 2001,[15] with its origins going all the way back to the late 1970s and my 1979 List.

Extracurricular affairs in Poland

My trips to Warsaw for joint work on GAS experiment typically lasted one or two weeks, with Grzedzielski's Space Research Center being a hospitable host (Fig. 9.7). The visits also opened a window to a new world. In Warsaw, one could see American movies that did not reach the Soviet Union.

---
14. Hertz and Wagner, 2003
15. Gruntman et al., 2001

9. ENA Experiment That Never Flew — Mike Gruntman

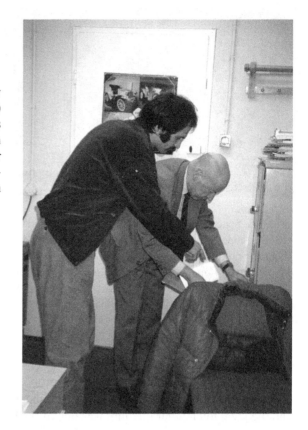

Fig. 9.7. Stanislaw Grzedzielski (right) and the author of this book (left) at the Polish Space Research Center in Warsaw in the mid-1980s. From collection of Mike Gruntman.

At a flea market, it was possible to find books by Orwell and Solzhenitsyn in English, banned in the Soviet Union. Polish newspapers also enjoyed more freedom than their counterparts in Moscow. Reading and speaking a little bit of Polish opened many doors.

When we ramped up work with the Warsaw colleagues, Poland was under martial law imposed by General Wojciech Jaruzelski in December 1981. Polish communists tried to preserve the regime and crush the existential threat from the *Solidarność* (*Solidarity*) trade union movement. Very quickly I befriended activists of the underground Solidarity in Warsaw, including those who had been arrested by the security forces on the night of the military takeover. The authorities released many of them after several months in detention camps.

When I was in Warsaw on one of my trips, Grzedzielski left for Moscow to attend the launch of the space mission Vega 1 to comet Halley. The

Fig. 9.8. Lapel pin made by the outlawed Solidarity operating underground in the early 1980s. The word *Solidarność* (Solidarity) was always printed in a slightly dark red color. From collection of Mike Gruntman.

spacecraft went into orbit on Saturday, December 15, 1984. On Sunday morning I took a train to the city of Gdansk on the Baltic coast for a day of sightseeing. It was a three-hour ride.

Gdansk was the birthplace and a major stronghold of the NSZZ Solidarność (Fig. 9.8). NSZZ stood for *Niezależny Samorządny Związek Zawodowy* (*Independent Self-Governing Trade Union* in Polish). My friends suggested arranging a brief meeting in Gdansk with the Solidarity leader Lech Walesa (Wałęsa), who had been released from prison two years earlier in November 1982. Walesa would later become the first, after World War II, freely elected Polish president, from 1990-1995.

Fortunately, I did not view myself as important and the meeting did not take place. It was a wise decision since the agents of state security and informers certainly saturated Walesa's surroundings and monitored his every step.[16] Chances were that I would have been flown from Warsaw, figuratively speaking, straight to Siberia rather than return to Moscow.

One can only speculate about a possible impact of such a not-unlikely career-ending turn of events on the advancement of ENA imaging, particularly on mapping of the interstellar boundary.

My sightseeing trip brought me to the right place at the right time. Or, depending on perspective, to the wrong place at the wrong time. That day, December 16, 1984, would witness major clashes between the anti-communist protesters and riot police using "tear gas, riot sticks, and smoke bombs ... to disperse supporters of the banned Solidarity... At least 12 people were reported detained... [And] several people in the crowd were beaten by policemen who charged into the throng."[17]

---

16. Declassified Polish state archives show today that some leaders of the Solidarity trade union and its active supporters collaborated under coercive pressure with the Polish Security Service. The degree of collaboration, duration, and other details remain murky.
17. Kaufman, 1984

9. ENA Experiment That Never Flew   Mike Gruntman

Only 1600 ft (500 m) away from the Gdansk railroad station, I ran into and joined a large crowd around St. Bridget's Church in the city center. Its priest and a staunch supporter of Solidarity, Father Henryk Jankowski, provided sanctuary to trade union leaders. Elements in the Catholic Church had been playing a major role in spiritual resistance to communists throughout Poland for decades. Underground activists often used the holy masses as starting points for political events.

On that December day, a few thousand people pressed to the entrance waiting for the mass to end. Then, several Solidarity leaders exited the church. I recognized Walesa. A person next to me pointed at another leader, Andrzej Gwiazda,[18] arrested in December 1981. Gwiazda spent more than two years in prison and was released on amnesty only a few months earlier.

The people began chanting anti-communist slogans and the names of the trade union leaders, especially of Walesa and Zbigniew Bujak.[19] Bujak avoided arrest in December 1981 and headed the outlawed Solidarity operating underground. The government would capture him only in 1986. The crowd periodically broke into chanting "*Nie ma wolności bez Solidarności*" ("There is no freedom without Solidarity [trade union]").

Then, we headed towards the Three Crosses (*Trzy Krzyże*) monument 2000 ft (600 m) away. I could say "we" because by that time I had become part of the protest, had turned into an active participant, and was marching with the gathered men and women.

The Three Crosses stood next to the entrance to the *Lenin Shipyard*, where the activists formed the Solidarity Trade Union in 1980. The monument commemorated the shipyard workers killed by the communist government forces during unrest 14 years ago in 1970.

The infamous brutal paramilitary police, ZOMO, blocked the passage to the Three Crosses. ZOMO stood for *Zmotoryzowane Odwody Milicji Obywatelskiej* in Polish, or *Motorized Reserves of the Citizens' Militia*. These units played the role of special forces of the police.

Clashes ensued, with the paramilitary police repeatedly charging the crowd and attacking people with batons and tear gas. For the first time in my life, I tasted tear gas in the middle of the scuffle. At one moment I was also close to getting hit by police batons. Security agents detained Gwiazda and he would spend the next five months in prison again.

Walesa tried to bring flowers to the monument, but the police stopped him. The *New York Times* described that

---

18. Before introduction of martial law in December 1981, Gwiazda served as vice president of Solidarity.
19. Bujak headed Solidarity in the Warsaw region (Mazovia Province) in 1981.

[w]ords were exchanged [with the police line] and witnesses said there was more elbowing and some shoving. Mr. Walesa, reportedly looking disgusted, put his flowers at the feet of the police and withdrew.[20]

The *Wall Street Journal* also summarized what happened on that day,

> Solidarity founder and Nobel peace laureate Walesa and about 3000 backers of the outlawed [Solidarity] union were marching to mark security forces' killing of shipyard workers 14 years ago in riots over food price rises. Police prevented the protesters from reaching a monument to the slain workers and detained over a dozen people. It was described as the worst clash in Poland in a year.[21]

This was wintertime, and most people wore clothing of dark colors. In contrast, I had my new synthetic winter jacket of whitish color (Fig. 9.9) which clearly stood out. I still remember that I then thought about officers of the Polish internal security service who would later review the photographs of the demonstration participants and try to identify that new, unfamiliar guy in a white jacket, shouting and raising his fist. Socialist countries were poor by Western standards, and people wore the same clothing for many years. So subsequently, my whitish winter jacket would get "involved" in another protest story several years later (Chapter 11).

In the early evening, I took a train back to Warsaw. The Gdansk contacts of my Solidarity friends knew my itinerary. They came to the railroad station to check on me after the clashes with the riot police and arrests. The new friends in Gdansk worried for my safety and were happy to see me in one piece and in good spirits. They brought me sandwiches for the return trip and waved goodbye. Good people.

It turned out that they were not the only ones who worried. When I walked to the Space Research Center on Monday morning, the director's secretary asked me to immediately see Grzedzielski. On his way home from the VEGA launch, Stanislaw heard the news about clashes with the ZOMO riot police in Gdansk. And he knew about my plans to go to Gdansk that day.

Many years later, I attended a meeting at Stanford University. I flew in early and spent several hours at *Hoover Institution*. Its library and archives hold probably the only comprehensive collection of *Mazovia Weekly* (*Tygodnik Mazowsze*) in the United States. During the period of martial law and thereafter, this was an influential underground publication by the banned Solidarity operating in Mazovia, the central region of Poland that

---

20. Kaufman, 1984
21. Polish police battled ..., 1984

9. ENA Experiment That Never Flew — Mike Gruntman

included Warsaw.

During one of my trips to Poland, I took my saved per diem paid by the USSR Academy of Sciences and passed it to the underground Solidarity. It was ironic and satisfying to give communist-state-provided money to an anti-Soviet movement. Every issue of *Tygodnik Mazowsze* published confirmations of the funds received by the outlawed trade union, usually listing a code name, a *nom de guerre*, of the donor and the amount of money. I believe that I found the cryptic entry confirming my contribution.

Fig. 9.9. The author of this book in his whitish synthetic winter jacket in Warsaw in the mid-1980s. This clothing stood out in the crowd of the supporters of the outlawed Solidarity dressed in dark colors during clashes with Polish riot paramilitary police in December 1984. The jacket would "participate" later in another protest in Moscow. From collection of Mike Gruntman.

Early advances of ENAs

Our early work contributed to the development of the first concepts of ENA imaging and associated instrumentation during my 15 years at IKI from 1975 till the end of the 1980s. We concentrated on imaging the interstellar boundary of the heliosphere in energetic neutral atom fluxes and measurements of the neutral solar wind and to a lesser extent on magnetospheric ENAs.

Particularly important were our advances in the physics of ENA detectors, which was especially consequential during those early stages. Without exploring various approaches and showing how ENAs could be measured, energetic neutral atoms would have remained an esoteric curiosity much longer. This pioneering effort also led to the building of dedicated first-generation ENA space instruments. While the joint Soviet-German-Polish GAS experiment never took off from the launch pad and reached space, it helped mature the concepts.

Fig. 9.10. Some early ENA space missions. Top: guest pass to launch of the IMAGE mission from the Vandenberg Air Force Base in 2010 (left) and a TWINS mission sticker (right). Bottom: stickers for space mission IBEX. From collection of Mike Gruntman.

9. ENA Experiment That Never Flew   Mike Gruntman

Many detector techniques, elements, components, and ideas (Fig. 7.6) that are used today in ENA instruments had been proposed, analyzed, developed, or demonstrated during the 1970s and 1980s. Other research groups across the world would advance them further later and apply them to the next generations of much more capable ENA instruments. The dedicated experiments would image the terrestrial magnetosphere (IMAGE, TWINS) and interstellar boundary (IBEX) in ENA fluxes after the year 2000 (Fig. 9.10).

Working relations with several colleagues in West Germany and Poland and later in the United States enabled our progress in the direct detection of neutral atoms. Much was achieved in an essentially joint effort. It was vitally important that we had established this mutually beneficial collaboration by the early 1980s. Without this propitious timing, many advances would not have been achieved at IKI. It was even possible that the entire ENA field would have then died and disappeared in the institute, lacking interest and support.

Several unrelated factors conspired and converged to advance ENA imaging. Most important among them were the opening of IKI to collaboration with the Western world by institute director Roald Sagdeev; the atrophied and increasingly malfunctioning Soviet system of total control; the self-interest of foreign scientists to take advantage of the opportunities of "free rides" of their instruments on Soviet science spacecraft; geopolitical developments, including pressure on the Soviet Union by U.S. President Reagan; the unique position of Poland and specifically Stan Grzedzielski, who effectively served as an interface between the East and West; and the personal quirks, interests, talents, flaws, and idiosyncrasies of the scientists involved.

10. Neutral Atoms Reach Critical Mass

Serendipitous discovery of magnetospheric ENAs

Realization in the early 1980s that ENAs offered a new window on magnetospheric processes boosted the entire neutral-atom field. The interest in ENA properties increased after the serendipitous discovery of energetic neutral atoms made by energetic particle instruments studying magnetospheres on several spacecraft. The scientists working in this field of space physics significantly outnumbered those focused on the heliosphere and the solar system interaction with the surrounding interstellar medium. Magnetospheric physics also directly addressed the practical needs of spaceflight because most of the spacecraft operated in the terrestrial magnetosphere and ionosphere.

Solid-state detectors constituted the basis of energetic particle instruments and often discriminated against electrons. Such instruments did not distinguish, however, between an ion and ENA. Consequently, an energetic particle detector served as an ENA detector in the absence of the normally abundant ions.

Voyager 1 reported possible detections of ENAs during flybys of the magnetospheres of Jupiter and Saturn in March 1979 and November 1980, respectively. Then magnetospheric ENAs were identified in measurements made by the IMP 7, IMP 8, and ISEE 1 spacecraft near the Earth (Appendix B). These measurements represented a milestone in the validation of the idea that global magnetospheric processes could be remotely studied by measuring ENAs.

In 1987, Edmond (Ed) Roelof (Fig. 10.1) from the Applied Physics Laboratory, APL, in Maryland reconstructed the first global ENA image

10. Neutral Atoms Reach Critical Mass   Mike Gruntman

Fig. 10.1. Edmond (Ed) Roelof of the Applied Physics Laboratory, Johns Hopkins University, at the Space Gas Dynamics Conference in Moscow in 1988. Photograph courtesy of Mike Gruntman.

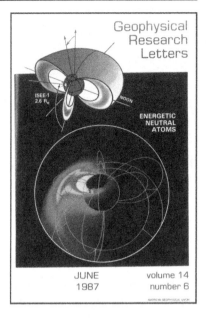

Fig. 10.2. Cover of *Geophysical Research Letters* in June 1987 highlighting the first reconstructed ENA image of the Earth's magnetosphere during a geomagnetic storm. It prominently displayed the words "Energetic Neutral Atoms." The issue featured a consequential article by Ed Roelof: "Energetic neutral atom image of a storm-time ring current." Cover credit: American Geophysical Union.

of the storm-time ring current on September 29, 1978, from the ISEE 1 data.[1] A leading science journal, *Geophysical Review Letters*, highlighted this consequential work on its cover, with the words "Energetic Neutral Atoms" prominently displayed (Fig. 10.2). Large influential space science groups and the space physics division at the NASA headquarters began to pay attention to ENAs.

Our main interest at IKI focused on the neutral solar wind, heliospheric ENAs, and interstellar helium. At the same time, we had also been looking

---

1. Roelof, 1987

into magnetosphere imaging since the late 1970s (Fig. 7.3). The acknowledgment section[2] in one of our publications, "Experimental opportunity of planetary magnetosphere imaging in energetic neutral atoms," reflected the emerging international links and interests in the ENA field in 1986 (Fig. 10.3). After obligatory thanks to our patron-saint department head academician Petrov, we acknowledged Grzedzielski (CBK, Poland); Axford, Rosenbauer, and Arne Richter (MPAe, West Germany); Don Williams and Andy Cheng (APL, USA); and Hsieh (University of Arizona, USA).

Fig. 10.3. Cover of a 1986 publication (Gruntman and Leonas, 1986) on magnetosphere ENA imaging (left) and its acknowledgment section (right).

International workshops on neutral atoms

The Warsaw Space Research Center played a critically important role in building an international community of scientists interested in neutral atoms in space. CBK Director Stan Grzedzielski maintained extensive working relations with neutral-atom German colleagues at the University of Bonn and MPAe in Katlenburg-Lindau. An excellent scientist, Stan attracted researchers in his center to work on various aspects of ENAs and heliospheric interactions. He was also a diplomat, recognized for his suave abilities to serve later as the executive director of COSPAR for several years.

With the support of the Polish Academy of Sciences, Grzedzielski initiated a series of annual international workshops focused on neutral atoms in space. These consequential gatherings helped consolidate and mature the "fragile" emerging field. The very first meetings primarily concentrated

---

2. Gruntman and Leonas, 1986, p. 25

10. Neutral Atoms Reach Critical Mass — Mike Gruntman

on neutral atoms in interplanetary space, reflecting the main interests of scientists at IKI, CBK, MPAe, and the University of Bonn. As time went by, the scope widened to include other related areas, particularly magnetospheric ENAs. Appendix A shows the scientific programs of the first four annual workshops from 1984-1987. Fortunately for preserving history, CBK included the scientific programs of the first three workshops in the materials of the fourth workshop,[3] distributed among its participants. The lists of speakers and the titles of their presentations illustrate the evolution and steady growth of the field.

The very first small meeting, "Detection of Neutral Gases in Interplanetary Space," took place in Warsaw in March 1984. Its participants included Fahr from Bonn; Rosenbauer and Witte (Fig. 8.12) from Katlenburg-Lindau; and Baranov, Leonas, and myself from Moscow. Grzedzielski brought several active researchers from his Space Research Center to this new field. Growing the neutral-atom community was particularly important during the early days.

The participation of young CBK scientists also reflected expanding work in Warsaw on our joint Soviet-German-Polish GAS experiment. The list of the speakers at the first several workshops[4] also showed that not a single other IKI staff member besides the author of this book participated in work on the GAS experiment in those years. This state of affairs reflected the priorities of IKI and Department No. 18.

CBK's Stanislaw Bleszynski (Stanisław Błeszyński) and Daniel Rucinski (Rućinski) engaged in the neutral-atom field under Grzedzielski's scientific direction and gave presentations at the first workshop. Bleszynski continued his work for some time, then left Poland in the 1990s, and finally quit science altogether. Rucinski (Fig. 10.4) would be making important contributions to studies of heliospheric neutrals for two decades. A good friend, Daniel tragically ended his life in the 2000s.

Another participant of the first workshop, CBK theoretician Wieslaw (Wiesław) Macek (Fig. 10.5), periodically published articles on neutral atoms. His main science interests concentrated, however, outside that field. Wieslaw also became a good friend. A Warsaw scientist, Marek Banaszkiewicz, brought attention in his joint talk with Grzedzielski to ENAs originating from Jupiter. He later worked with MPAe's Rosenbauer and Witte on data reduction of the interstellar helium experiment on Ulysses. Marek then served as director of the Space Research Center in Warsaw in the mid-2000s.

---

3. Bzowski et al., 1987
4. Appendix A; Grzedzielski and Page, 1990

Fig. 10.4. Daniel Rucinski of the Space Research Center in Warsaw has played an active role in the study of neutral atoms in space since the early 1980s. Photograph (1988) courtesy of Mike Gruntman.

Fig. 10.5. Wieslaw Macek of the Space Research Center in Warsaw published on neutral atoms in space in the 1980s and 1990s. Photograph (the mid-1990s) courtesy of Mike Gruntman.

For many years, Hans Fahr had been able to secure financial support for collaborative projects with Polish and later Soviet and then Russian scientists (Fig. 10.6). He invited them to spend some time at the University of Bonn for joint theoretical work, which helped the colleagues in Warsaw in the 1980s and Moscow in the 1990s financially, especially during hard economic times in their respective countries. Even during "normal" economic periods, the life of young rank-and-file scientists under communist rule and during early transitional years afterward was hard economically. Their visits to Bonn helped neutral-atom-related science continue.

Baranov wrote in his memoirs that Fahr "did a lot for our [Moscow heliosphere theoreticians] survival during the time of 'perestroika.'"[5] I am sure that Grzedzielski would share this assessment of help provided to the Polish group during the 1980s and early 1990s. Hans Fahr played a truly consequential role in maintaining and growing heliospheric research, in-

---

5. Baranov, 2009, p. 822

10. Neutral Atoms Reach Critical Mass — Mike Gruntman

Fig. 10.6. Stan Grzedzielski (left) and Hans Fahr (right) in 2005. For many years Fahr supported joint projects with Polish and Soviet and then Russian scientists, providing a financial lifeline for many theoreticians during the days of economic hardships behind the Iron Curtain in Eastern Europe. Photograph from collection of Mike Gruntman.

cluding neutral atoms, beyond the West German borders in divided Europe at that time.

Working on the experimental side and verboten to travel, I did not have an opportunity to spend time in Bonn for joint work. However, my frequent communications and occasional meetings with Hans provided major stimulation in an exploration of the processes in the heliosphere. From time to time, I came up with interesting, as they looked to me, theoretical questions and ideas. After obtaining physics estimates, I looked up whether something had already been published on the subjects in the scientific literature. More than once I found that Hans had considered similar phenomena. Our collaboration and friendship continue to this day (Fig. 10.7).

Fig. 10.7. The author of this book (left) with Hans Fahr (right) in Ahr Valley in Germany 30 years after their first meeting in Moscow. Photograph (2009) from collection of Mike Gruntman.

10. Neutral Atoms Reach Critical Mass — Mike Gruntman

Grzedzielski's Space Research Center organized the second annual workshop "Problems of Physics of Neutral Particles in the Solar System" in Zakopane in southern Poland in September 1985. Many participants of the first meeting took part. There were additions as well. A few more Polish scientists joined the effort, particularly Marek Rubel, who began experimental studies of sputtering related to solar wind bombardment of interplanetary dust.

Fig. 10.8. Maciej Bzowski (left) and Romana Ratkiewicz (right) of the Space Research Center in Warsaw have been working on the heliosphere and neutral atoms since the mid-1980s. Photograph from collection of Mike Gruntman.

Another new CBK participant was Maciek Bzowski (Fig. 10.8), who explored various processes in the heliosphere. He later became a co-investigator of NASA's ENA-imaging IBEX mission in the mid-2000s. Today, Bzowski leads a group of the next generation of heliospheric scientists in Warsaw. A few of his younger staff members also transferred to the ENA program of IBEX Principal Investigator Dave McComas at Princeton. I joked with Maciek that his office in the CBK building must be on a very high floor since his young collaborators could "see" New Jersey shores from there.

Fig. 10.9. Guenter Lay, a long-time close collaborator of Hans Fahr at the University of Bonn, worked on sounding-rocket optical experiments to study neutrals produced by interplanetary dust. One such project, Zodiak, attempted to detect hydrogen atoms outgassing from dust grains in the Sun's vicinity. Photograph (2005) from collection of Mike Gruntman.

Guenter Lay (Fig. 10.9) from Fahr's group in Bonn gave a talk on their experimental effort to optically observe neutral atoms near the Sun from sounding rockets. A few East German scientists also attended the workshop in Zakopane.

Nobody came to Zakopane that year from Moscow. My boss Leonas did not let me go because of our periodic internal frictions. He and Baranov chose not to attend as well. Chasing elusive individual neutral atoms in space did not exactly fall into their primary science interests. They also did not go to the third annual workshop in Guenzburg nor far from Ulm between Munich and Stuttgart in West Germany, in September 1986. For me, this meeting in Bavaria was obviously out of the question because it took place beyond the barbed-wire perimeter of the socialist camp.

This latter meeting "Plasma-Gas Interaction in Space" in Guenzburg witnessed a significant increase in the number of participants, especially from West Germany. CBK's theoreticians Joanna Ziemkiewicz and Romana (Roma) Ratkiewicz also joined the workshop. Roma (Fig. 10.8) focused on theoretical modeling and numeric simulations of the heliosphere interaction with LISM. She continues this work today.

In addition, for the first time, several scientists from the United States attended. For them, travel to FRG, a NATO country, was much easier than to communist Poland during the Cold War. The list of workshop participants included Darrell Judge from the University of Southern California, Ed Roelof and Don Mitchell from the Applied Physics Laboratory, and Joe Ajello from NASA's Jet Propulsion Laboratory. The field of neutral atoms was steadily approaching critical mass.

What is 'Whoa'?

The next, fourth annual workshop on "Interaction of Neutral Gases with Plasma in Space" took place again in Poland in September 1987 (Fig. 10.10). The Polish group organized it in Radziejowice about 25 miles (40 km) southwest of Warsaw.

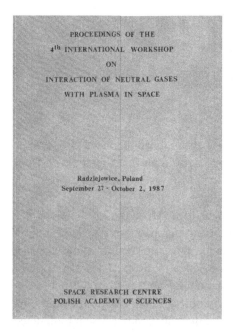

Fig. 10.10. Cover of the Proceedings of the 4th International Workshop on *Interaction of Neutral Gases with Plasma in Space*, 1987 (Bzowski et al., 1987).

In addition to the "usual" strong participation of German scientists, a few Americans also came to Poland for the first time, including Johnny Hsieh from the University of Arizona and Darrel Judge (Fig. 10.11) and Howard Ogawa (Fig. 10.12) from the University of Southern California. Baranov, Leonas, Malama, and I represented the Moscow group at the workshop. Also, a collaborator of Baranov from the Moscow State University, theoretician Mikhail G. Lebedev, joined the team.

The Radziejowice workshop took place in a secluded restored palace belonging to the Polish Academy of Sciences. In the past, it served as a residence for high nobility and royals and now housed the meeting rooms and living quarters of the participants. After presentations during the day and dinner in the evening, everybody congregated to a common area with adult beverages and snacks for a continuation of discussions on neutral atoms and other related and unrelated topics.

Darrell Judge (Fig. 10.11) from USC and I had never met before but we knew of each other by publications. On the first night, I decided to show

Fig. 10.11. Top: Darrell Judge (left) and Stan Grzedzielski (right) at the fifth annual international neutral-atom workshop, Space Gas Dynamics Conference, in Moscow in October 1988. Peter Blum, University of Bonn, is in the background (center). Photograph courtesy of Mike Gruntman.

Bottom: Two years later in December 1990, Stan Grzedzielski and his wife Bozena (center) at the home of Darrell Judge (right) in Los Angeles. The author of this book, after his "tunneling transition" to California, is on the left. Photograph from collection of Mike Gruntman.

10. Neutral Atoms Reach Critical Mass   Mike Gruntman

Fig. 10.12. The author of this book (left) with Howard Ogawa (right) at his office at the University of Southern California in 1991. Photograph from collection of Mike Gruntman.

my "knowledge" of Americana to him. With a glass in hand, I approached Judge: "Darrell, do you know what Governor of North Carolina said to Governor of South Carolina when they met?"

With a twinkle in his eye, he questioned, "What did he say?" I replied, "Let's have a drink!" We laughed and drank. Then, he, in turn, asked me what Paul Revere said when he reached Lexington. I knew about Paul Revere and his famous ride, but I did not know the answer. So, Darrell said, "Whoa!"

Never shy to say that I did not know something, I sincerely questioned, "What is 'Whoa'?" He explained that it was a command to a horse to stop or slow down. We laughed again and had a drink. This became my introduction to Darrell's life-long love of horses. Later in life, he would have half a dozen horses at his ranch 100 miles southeast of Los Angeles. We became instant friends and then colleagues at USC (Fig. 10.11), as described in Chapter 11.

Now USC's Howard Ogawa posed his "burning" question. He wanted to know whether it was true that many people in Eastern Europe and Russia could drink more than a shot of hard liquor in one gulp. I poured two or three ounces of whiskey in a glass and proficiently demonstrated that basic Eastern European qualification to him. Amusingly, Howard's fascination did not have bounds. He would be recalling that episode for many years to come. We became friends and later colleagues at USC as well (Fig. 10.12).

Fifth Workshop in Moscow

We organized the fifth annual workshop on neutral atoms, called the "Space Gas Dynamics Conference," in Moscow on October 10-14, 1988. IKI's Department No. 18 had transferred by that time to the Institute for Problems in Mechanics of the USSR Academy of Sciences (Chapter 11). The conference badges prominently showed the Roman numeral V in the background, referring to the fifth annual meeting (Fig. 10.13).

This successful conference attracted 75 participants and included more than 50 talks. In addition to energetic neutral atoms, the speakers addressed various gasdynamical aspects of the interaction of the heliosphere with the LISM and the solar wind with comets, reflecting the thrust of theoretical work by Vladimir Baranov and his associates. Some participants even discussed interactions of stellar winds in binary stellar systems.

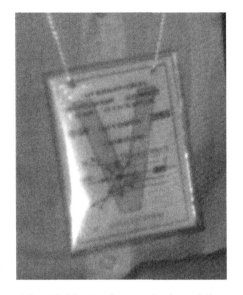

Fig. 10.13. Conference badge of the fifth neutral atom workshop, Space Gas Dynamics Conference, in Moscow with the Roman numeral V prominently in the background. Photograph from collection of Mike Gruntman.

The workshop provided another opportunity to expand scientific connections with colleagues around the world. In total, 28 foreign scientists from Bulgaria, France, East Germany, West Germany, Great Britain, Finland, Japan, Poland, and United States came to Moscow.[6]

New participants included Eberhard Moebius (Fig. 10.14), who had detected pickup ions in the solar wind that originated from the interstellar neutral atoms penetrating the heliosphere. Studies of this distinct ion population opened an important window into the composition of the interstellar gas surrounding the solar system. Eberhard would soon move to the United States from the Max Planck Institute in Garching near Munich.

Ten years after the Moscow conference, Eberhard Moebius, George Gloeckler, Hans Fahr, and I made the first step in formulating a concept

---

6. Baranov and Leonas, 1989

10. Neutral Atoms Reach Critical Mass — Mike Gruntman

of a space mission, "A Local Interstellar Medium Explorer (LIME),"[7] focused on the interaction of the heliosphere with the local interstellar medium. The acronym did not survive but the effort began an indispensable process of programmatic advancing and firming up an idea of a dedicated heliospheric ENA mission. It would ultimately lead to proposals (Fig. 9.6) culminating in IBEX.

Another new conference participant, Alan Lazarus from the Massachusetts Institute of Technology, conducted important observations of the distant solar wind on the Voyager spacecraft. Staff scientists at the *Aeronomical Service (Service d'Aéronomie)* in France, Jean-Loup Bertaux and Rosine Lallement, presented their work on optical observations of interplanetary neutral atoms. Bertaux, Lallement, and I had met earlier on one of their visits to IKI.

The Moscow workshop allowed me to meet in person for the first time Ed Roelof from APL. His recently reconstructed ENA image (Fig. 10.2) of the storm-time ring current in the Earth's magnetosphere had attracted attention to the entire ENA field. Roelof became a close colleague and collaborator in ENA imaging for many years to come. Our scientific cooperation and friendship continue to this day (Fig. 10.15).

Fig. 10.14. Eberhard Moebius at the international Space Gas Dynamics Conference in Moscow in October 1988. Photograph courtesy of Mike Gruntman.

The Moscow conference led to serious internal problems and an unnecessary fissure in the Moscow group. Heads of administrative units Baranov and Leonas were the primary organizers of the event. Baranov later wrote that he could not help but

> admire the high level of responsibility, sacrifice, selflessness, which were shown by all members of my [Baranov's theoretical] laboratory as well as young talented physicist-experimentalist Misha Gruntman [in organizing the Moscow neutral-atom meeting].[8]

---

7. Moebius et al., 1998
8. Baranov, 2009, p. 362

Fig. 10.15. Ed Roelof of the Applied Physics Laboratory attending an IBEX team meeting in Bern, Switzerland, in 2007. Photograph courtesy of Mike Gruntman.

Baranov correctly noted the huge effort and sacrifice of the organizers. He identified only me by name among the scientists in the quoted passage. In fact, most other scientists and engineers of the department contributed very little to hosting the event. Baranov and Leonas assigned an insufficient number of scientific and engineering staff to support the conference. They ignored my protests, and numerous logistical problems, much more than a fair share, fell on me.

I was in disbelief and truly appalled by what happened. Our relations and trust would never recover and be the same. It took a special effort to keep this episode from negatively affecting our professional activities and work.

10. Neutral Atoms Reach Critical Mass   Mike Gruntman

Trip to the Caucasus

On a positive note, I succeeded in organizing a smooth trip for workshop participants Blum, Fahr, Grzedzielski, and Rosenbauer as guests of the Academy of Sciences to the Special Astrophysical Observatory and Byurakan Astrophysical Observatory after the meeting. My ongoing collaboration on positions-sensitive detectors with scientists in these two institutions made us truly welcome there, not just a burden imposed by the faraway Academy officialdom.

We first flew to a regional airport hub, Mineralnye Vody, in the North Caucasus and then were driven to SAO (Fig. 10.16). The highlights included visits to the then largest in the world 6-meter optical telescope (Figs. 5.9, 5.11) and the 600-m radiotelescope RATAN-600 (Figs. 5.13, 5.14, 6.4). Nearby pristine mountains provided opportunities for informal interactions with the hosts in nice, crisp October weather (Figs. 6.4 and 10.17).

Fig. 10.16. Tasting local kebabs on the way to the Special Astrophysical Observatory from a regional airport hub, Mineralnye Vody, in the North Caucasus in October 1988. Right to left: the author of this book, Hans Fahr, Stan Grzedzielski, a local man, Peter Blum. Photograph (October 1988) credit: Helmut Rosenbauer.

Fig. 10.17. Can one see the heliopause from there? From top to bottom: the author of this book, Hans Fahr, Stan Grzedzielski, and an accompanying engineer from the Baranov's group. Photograph (October 1988) credit: Helmut Rosenbauer.

Then we flew to Yerevan, the capital of Armenia, and visited the nearby Byurakan Astrophysical Observatory. The region had already plunged into turmoil, after the recent pogrom in the town of Sumgait near Baku in neighboring Azerbaijan. Seven months earlier in February 1988, a total of more than 30 people, primarily Armenians but also some Azerbaijanis (Azeris), were killed there. Violence in the contested *Nagorno-Karabakh* region, also known and dear to the Armenians as *Artsakh*, intensified. Mutual ethnic cleansing followed, with thousands of miserable refugees.

I visited the region only six weeks earlier when my vacation time came up in early September of that year. I went first to Baku and Sumgait (Fig. 10.18) in Azerbaijan, then Stepanakert (Fig. 10.19) and Shusha (Shushi) in Nagorno-Karabakh, and finally reached Yerevan. In Shusha I could have gotten killed if caught when photographing local Azeris using an Armenian cemetery as a pasture for their sheep.

Fig. 10.18. Orwellian-named *Monument to Friendship* (as in friendship among the peoples) on a square in Sumgait, a dreary industrial town 20 miles (30 km) from the Azerbaijan capital Baku. Mobs killed more than 30 people, primarily Armenians, in a pogrom in late February 1988. The Armenians constituted 7% of the town population at that time. Photograph (September 1988) courtesy of Mike Gruntman.

Fig. 10.19. Memorial plate "To Victims of Sumgait Tragedy" in Stepanakert, the capital of the Nagorno-Karabakh (Artsakh) region in September 1988. The inscription is in Russian, the lingua franca of the Soviet Empire. Photograph courtesy of Mike Gruntman.

My almost empty tourist hotel in Shusha housed a small group of stranded prosecutors. The government sent them from central Russia to help with law enforcement, which had completely collapsed by that time. The prosecutors lived isolated in the hotel for a month with not much to do except watching the unfolding events. The local authorities disintegrated and split along ethnic allegiances. They were happy to see a new face, especially from Moscow, and eager for news and chat. Over the drinks, they confirmed that my encounter earlier that day at the cemetery had been a very close call. What saved my life was that the local Azeris did not see me throwing my compact camera in the tall grass when approached.

In Yerevan, my local science friends introduced me to a couple of members of the *Karabakh Committee* that organized massive protests in the streets (Fig. 10.20, top left). People demanded the unification of Artsakh with Armenia, chanting *"Artsakh miatsum!"* ("Unite with Artsakh!" in Armenian). The Soviet authorities arrested the leading Committee members three months later. In a few years, some of them would become members of the government of independent Armenia.

An area surrounding the Opera Theater became the central place for mass protests in Yerevan (Fig. 10.20). I mixed with a crowd there for half an hour when a member of the Karabakh Committee found me and pulled me aside. He sternly advised me to keep my mouth shut. Somebody had told him that I had been explaining in conversations with the gathered people that their problem was not with the personal unfriendliness and incompetence of Soviet leader Mikhail Gorbachev but with the Soviet system. "Look, local KGB officers know us all and we know them," my new friend said to me. "They would not touch us, the Armenians, as we are all here related, the cousins, brothers, uncles, and members of same families. But they will break your neck, the outsider." It is even possible that an undercover security officer had tipped off my protector from the Karabakh Committee. I was an *odar* in that land.[9]

When the neutral-atom workshop participants came to Yerevan one month later in October 1988, the subdued mood in the streets did not resemble the boisterous meetings of people (Fig. 10.20, top left) and mass demonstrations by excited students in September. Dispirited men and women still gathered at the square near the iconic Opera Theater to express their grievances (Fig. 10.20, bottom left). Posters blamed the Moscow rulers and personally Gorbachev for allowing the killing of the people to happen. They demanded the authorities recognize the Sumgait massacre as

---

9. *Odar* (pronounced o-DAR in Armenian) means a *non-Armenian*

10. Neutral Atoms Reach Critical Mass — Mike Gruntman

Fig. 10.20. The subdued mood in the square near the iconic Opera Theater in Yerevan (bottom left) during the trip of the Space Gas Dynamics Conference participants in October 1988. The area continued to be the gathering place for people to express their grievances and show dismay with government actions and inaction. The banner on the theater building reads "sovereign government is people's will" in Armenian.

The same square boiled with thousands of angry people (top left) a month earlier, when I was there during my vacation trip on September 14. They chanted "*Artsakh miatsum!*" ("Unite with Artsakh!"), challenging the Moscow authorities over their handling of ethnic violence and cleansing in a conflict with Azerbaijan.

Right: photographs of the Armenians murdered in the Sumgait pogrom in February 1988. The banner above reads (in Russian) "The Supreme Soviet [Council of the USSR] must recognize genocide in Sumgait!"

The conflict was dormant for many years after the mid-1990s. In 2020, large-scale hostilities between the two independent countries, Azerbaijan and Armenia, erupted again. The rebuilt and strengthened Azeri military defeated the Armenian forces in Nagorno-Karabakh in a short campaign. Azerbaijan widely employed the new Turkey-built unmanned aerial systems with devastating effect.

Photographs courtesy of Mike Gruntman.

genocide (Fig. 10.20, right). The situation was gradually and irreversibly getting out of control.

The atmosphere differed in a restaurant only a few blocks away in the city center where we went for dinner. There, wine was flowing, accompanied by abundant and delicious food. The band played music and the guests danced. A party at one table across the hall recognized the foreigners and sent us a present, a bottle of sparkling wine commonly labeled champagne in the USSR. The local Armenians showed their hospitality to the guests of their city in a centuries-old tradition. Then, a waiter brought us another bottle of champagne, an expensive item on the wine list. These hospitable locals most likely operated in the informal economy or were perhaps criminal bosses. Their look supported such possibilities.

Stan Grzedzielski had known the Soviet Union very well and traveled extensively throughout the country. He observed at the end of our trip, "The Empire is still there. As guests and representatives of Moscow, we are still getting the attention of the local authorities. But clearly, the Empire is crumbling."

Sixth Annual Meeting: COSPAR Colloquium

The next year, 1989, witnessed the field of neutral atoms reaching critical mass. Grzedzielski organized the sixth annual meeting again in Poland, at a conference center in the town of Serock about 20 miles (30 km) north of Warsaw. This time, the gathering was under the aegis of COSPAR, the First COSPAR Colloquium "Physics of the Outer Heliosphere," September 19-22, 1989. Pergamon Press published the proceedings[10] of the meeting.

The colloquium attracted one hundred scientists (Fig. 10.21) who gave more than 70 presentations. The participants included many pioneers of the neutral atom field of the 1980s who had taken part in the earlier annual workshops. A number of specialists in related areas of science also came to the colloquium. A young researcher from a magnetospheric group at IKI, Stas Barabash, appeared at the neutral-atom meeting for the first time. Within several years, Stas would settle at the *Swedish Institute of Space Physics* in Kiruna and embark on building ENA instruments with Rickard Lundin. With time, Stas became director of the institute.

The once esoteric field of neutral atoms in space had matured and become mainstream. The interest of large and influential space science groups across the world was awakened, which would soon lead to flight programs. Within one decade, the dedicated ENA instruments, or ENA channels in

---

10. Grzedzielski and Page, 1990

10. Neutral Atoms Reach Critical Mass — Mike Gruntman

Fig. 10.21. Participants of the First COSPAR Colloquium "Physics of the Outer Heliosphere" in Serock near Warsaw in September 1989. Photograph credit: Space Research Center, Polish Academy of Sciences. Photograph credit: Space Research Center, Warsaw; from collection of Mike Gruntman.

charged particle instruments, would operate on or be readied for many space missions such as Ulysses, CRRES, POLAR, ASTRID, GEOTAIL, Cassini, IMAGE, TWINS, and others.

Energetic neutral atoms had reached critical mass.

11. Separation of Stages in Powered Ascent

When a space launcher reaches orbit, several mission-critical events occur during the powered ascent. The engine of the first stage cuts off, followed by firing explosive bolts connecting it to the second stage. Then, after a brief time interval for the first stage to separate, the second stage engine is ignited and the ascent continues.

Such "mission-critical events" happened on my personal trajectory in the late 1980s when the first stage of my life in the Soviet Union and work at IKI were approaching the cutoff, with the second stage preparing for separation and ignition.

Annual evaluation

Frictions between leaders of Department No. 18 of IKI and forceful institute director Roald Sagdeev reached a critical point sometime in 1986. The untenable state of affairs became especially clear during a routine annual evaluation of research staff.

That year, the director appointed a usual evaluation committee consisting of senior scientists of the institute as well as, ex officio, representatives of IKI's Communist Party Committee and the rubber-stamp trade union organization. My science performance exceeded expectations and I served as the instrument lead for the development of the joint Soviet-German-Polish flight GAS experiment.

The department's scientists entered the room individually, one after another, where the committee members considered their performance during the last year. The committee evaluated positively all staff scientists and engineers. However, it formally listed one specific flaw in the performance of everybody without exception, most likely on a direct order by Sagdeev

11. Separation of Stages in Powered Ascent — Mike Gruntman

or one of his deputies. All evaluations stated insufficiency of good scientific performance and called for "more involvement in space experiments."

My turn came in the afternoon. The committee chairman announced to me their determination of the need for more involvement in space experiments. I politely objected and pointed out the obvious contradiction. A rank-and-file research scientist at IKI could not be more involved in space experiments than being an instrument development lead, a position of major responsibility. The committee had praised my performance in the latter capacity just a few moments earlier. Several members of the committee also knew me personally and were familiar with my leading role in advancing the fields of position-sensitive detectors and energetic neutral atoms.

An awkward silence followed. The committee's "finding" and my "flaw" were obviously sheer nonsense, driven by internal politics. Everybody present perfectly knew and understood it. Nevertheless, the members unanimously confirmed their verdict. I only sarcastically said, "Wow!" and politely left the room. Silently to myself, I uttered a more colorful word, *le mot de Cambronne*.

It was truly sad to see some of the most accomplished and respected space scientists of the country acting in such a dishonest, slavishly obedient, and shameful way. It was a disgrace.

After receiving my evaluation, I immediately told my boss Leonas that I would go to see IKI director Sagdeev the next day. Knowing that I would certainly do this and expecting "fireworks," Leonas asked me to wait a couple of days. Two days later, he told me that our department head academician Petrov met with Sagdeev. The director deleted the "flaw" from my evaluation and from that of our electronics engineer, Boris Zubkov, who had been actively engaged in building another flight instrument for a space mission.

Transfer to IPM

This annual evaluation served as an important warning that the current modus operandi was not sustainable. A few months later the Academy struck an agreement to move the entire Department No. 18 from IKI to another institute of the Academy of Sciences, the *Institute for Problems in Mechanics*, IPM, a few miles away. IPM director, academician Alexander Yu. Ishlinsky, was an old friend of Petrov.[1] The transfer formally took place on January 1, 1987.

---

1. Today, the institute's name is *A. Ishlinsky Institute for Problems in Mechanics* of the Russian Academy of Sciences.

The time had come to part ways with the institute where I'd spent all my professional life, 15 years from 1973 till the late 1980s (Fig. 11.1). This began a transition phase in my life, a journey that had taken me from a child growing up at the Tyuratam missile test range[2] (also known as the Baikonur cosmodrome) to a staff research fellow at the leading space science institute in the country. The unfolding events unmistakably pointed to a new, yet unknown next stage of my life's trajectory.

Fig. 11.1. IKI "history wall" showing the institute's accomplishments and leadership from its founding in 1965. The author of this book stands next to the wall in 2016. The first stage of his space journey began as a child growing up at the Tyuratam missile test site in Kazakhstan (insert on the left, ca. 1958). He joined IKI as an 18-year-old student in 1973 and worked in the institute for 15 years till the late 1980s. Photographs from collection of Mike Gruntman.

Sagdeev made the relocation to the new institute painless. He allowed our experimental group to stay at IKI for more than one year while we prepared the new laboratory space and gradually transported and installed equipment at the new premises. I maintained my badge and continued to spend much of my time at IKI until mid-1989. This was a convenient place to meet with foreign colleagues as well. Besides, the IKI library had an excellent collection of space physics-related journals and other publications.

---

2. Gruntman, 2004, pp. 313-315; Gruntman, 2019; Smith, 2021

11. Separation of Stages in Powered Ascent   Mike Gruntman

The news about the transfer of Department No. 18 propagated like wildfire through the institute. The heads of three major science units of IKI immediately proposed to me, either over the telephone or in private meetings, to join their teams and stay in the institute. The boss of one other large unit diplomatically, not to antagonize Leonas and Baranov, let me quietly know that he would love to take me in.

These flattering offers meant an excellent assessment of my work and brought confidence. Nevertheless, I decided to remain in the group of Leonas and transferred to IPM with him. I do not know whether it was the right or best solution. Personal loyalty, justifiable or not, played a role.

On the other hand, I felt a premonition of bigger changes in life coming. As with a space launcher in powered ascent, the main engine of the first stage would soon be cut off, explosive bolts activated, and the engine of the second stage ignited, propelling me along the new trajectory. What I did not know was when, what, and how, and whether I would make it through this "mission-critical event" in one piece.

Whitish winter jacket goes from Gdansk to Moscow

The country was gradually turning unstable and authorities at all levels were loosening control. Frictions among various ethnic groups grew across the vast empire. Major clashes erupted in Kazakhstan in December 1986. Then, the conflict between Armenia and Azerbaijan became bloody and turned into an existential fight between the two peoples in 1988 (Chapter 10). Even such a politically reliable and docile part of Soviet society as the bulk of scientists in the Academy of Sciences began showing dissatisfaction and to ask heretofore forbidden questions. (The number of those with truly dissident views, in "internal emigration," and resisting the regime was always small.)

One day in late February 1989, a harried scientist from the laboratory of Gringauz found me in a reading hall of the IKI library. He said, "Rosenbauer has just arrived and he is looking for you. It is urgent."

In a few minutes, I saw beaming Helmut, who immediately asked, "Are you OK? Are you in trouble?" I could not understand what happened and what was this about. He explained that in early February, his wife, Hildburg, who had known me well, saw me in a news program on the German television "enticing the crowd." The reality was less dramatic.

Communist reformer Gorbachev regally allowed various entities, including the Academy of Sciences, to propose candidates for the election of deputies of the Supreme Soviet, a controlled and choreographed rubber-stamp parliament. The arrangement resembled choosing subservient representatives of the estates of the Marxist realm.

Consequently, in late January 1989, the politically obedient officialdom of the Academy of Sciences made public the list of trusted loyal academicians and rejected a few candidates for election proposed by the awakening Academy's rank and file. The eliminated candidates included dissident academician Andrei Sakharov and IKI director Sagdeev. Baranov, who got involved in the process as an "elector," representing IPM scientists, later described the Academy as "a very dark reactionary organization."[3]

Then, an event took place that would have been unheard of in the past. Grassroots activists organized a protest meeting against the manipulation of the candidate selection process by the officials of the Academy's Presidium. Scientists from fifty academic institutes gathered on February 2, 1989, near the buildings of the Presidium.

I was not involved in these activities but had to drop off some paperwork at Academy's offices. So I timed my trip in such a way as to take a look at the protest (Fig. 11.2, middle).

A couple of thousand scientists gathered with banners denouncing and shaming the Presidium of the Academy of Sciences. Some displayed the names of their institutes, including IKI (Fig. 11.2, top). Andrei Sakharov also attended the meeting (Fig. 11.2, bottom). Several foreign television crews were present. It was a peaceful gathering with practically no police around.

For many years, the Soviet people had been seeing on television screens footage of protests by "progressives" and other assorted socialists in foreign countries, with people chanting slogans and waving their hands. Nobody had experience in doing this in the totalitarian Soviet Union, where the Communist Party organized and staged all pro-regime meetings and demonstrations. The inertia of fear lived in the majority of the population. Repressive and coercive socialism ultimately destroys human dignity and corrupts souls.

Several activists and organizers spoke to the crowd. The meeting passed a resolution of no confidence in the Presidium and adopted appeals to the Academy of Sciences and the scientific community, demanding the Presidium to change its decisions on the selection of the candidates.[4] Some speakers shouted slogans, such as "Bureaucrats out!" or similar, and tried to excite the people.

The gathered scientists, probably one-half of them with Ph.D. degrees, remained static and unmoved (Fig. 11.3). The totalitarian Marxist state ef-

3. Baranov, 2009, p. 370
4. Baranov, 2009, p. 371

11. Separation of Stages in Powered Ascent   Mike Gruntman

Fig. 11.2. Meeting near the Presidium of the USSR Academy of Sciences in Moscow on February 2, 1989. Top and middle: scientists of the Academy with banners "New Statute to the Academy," "Out Academy bureaucrats!" and "Fire [sack] the Presidium of the Academy!" Other banners display the names of the institutes, including the Space Research Institute. Bottom: Andrei Sakharov at the meeting near television crews. Photographs courtesy of Mike Gruntman.

Fig. 11.3. Meeting near the Presidium of the USSR Academy of Sciences in Moscow on February 2, 1989. The scientists remained largely static and unmoved during the speeches of the activists. Photograph courtesy of Mike Gruntman.

ficiently eliminated the concept of a public protest. The meeting participants simply did not know what to do. For many, it was the first time in their lives that they defied the authorities, overcame fear, and came together to express their views. Resurrecting the civil society that was annihilated by the communists would take a few generations.

Somehow I ended up in the middle of the crowd. When a speaker again shouted a slogan, "Shame to the Academy," I raised my fist and yelled "Ya!" As the speaker followed with another slogan, I again raised my fist and shouted. After four or five such consecutive yells, the people around began joining me in raising their fists and shouting.

I noticed that whenever a speaker shouted a slogan, television cameras turned away from him to point at me in anticipation as I raised my fist and yelled. This pattern continued for a couple of minutes with more and more people shouting with me and the cameras getting back and forth.

Rosenbauer's wife saw this footage on German television. They got understandably worried about the consequences for me.

When I walked to IKI the next morning the first thing I saw was a page from a major national daily newspaper, *Sotsialisticheskaya Industriya*

11. Separation of Stages in Powered Ascent   Mike Gruntman

Fig. 11.4. Photograph in a major national newspaper (Scientists protesting, 1989) showing the author with a raised fist at a meeting near the Presidium of the USSR Academy of Sciences in Moscow on February 2, 1989.

The banners read "We are also the Academy" and "Sakharov, Sagdeev, Likhachev to deputies."

(*Socialist Industry*), pinned on a bulletin board near the entrance gate.[5] It prominently had my photograph (Fig. 11.4) with the raised fist in the middle of the protesters. I was again in the same whitish winter jacket as in Gdansk in 1984.

The centrifugal forces continued to grow across the country. Aside from the bloody conflict in the Caucasus and growing interethnic tensions elsewhere, almost two million people in the Baltic republics literally joined their hands in a human chain, stretching more than 400 miles (650 km). This action on August 23, 1989, called *The Baltic Way* (Fig. 11.5), commemorated the fiftieth anniversary of the Molotov-Ribbentrop pact of 1939 that cleared the way for the Soviet Union to "swallow" Estonia, Latvia, and Lithuania.

The "eternal" socialist paradise showed growing fissures. At the same time, the 19-million strong Communist Party of the Soviet Union still ruled the country and controlled the society and lives of the people. The Marx-

---

5. Scientists protesting, 1989

Fig. 11.5. Lapel pin of *The Baltic Way* action on August 23, 1989. Almost two million people joined hands to form a human chain stretching from north in Estonia through Latvia to south in Lithuania. From collection of Mike Gruntman.

ists, fellow travelers, left-leaning liberals, and other Soviet supporters in the West had mixed feelings about the events. Many adored Gorbachev and hoped for a return of the tarnished socialism to prominence. A keen observer, dissident Yuri Orlov, viewed it as "the curious phenomenon of Western Gorbamania" when "millions of people were riveted by the man they saw as the liberal czar of a backward nation."[6]

Insightful U.S. commentator Rush Limbaugh noted that "Gorbachev was genuinely thought to be savior" and "every time he and [U.S. President] Reagan got together [for a summit meeting], there literally were Gorbasms" by the leftists.[7] On the other hand, the weakening of the communist state by Gorbachev's policies of perestroika and glasnost dismayed numerous Western Marxists and dispirited many illiberal liberals. They cheered and rooted for the continuation of this cruel social experiment with an incalculable human cost.

Mission-critical event: separation of rocket stages

The times were changing. My relatively short "new life" at IPM and other related events are beyond the scope of this story.

I am not prepared yet to describe how exactly the engine of the first stage of my space launcher cut off, how the explosive bolts were activated, and how the stages separated. In rocketry, it is usually a short process, only a few seconds long (Fig. 11.6) but requiring a long, meticulous preparation. It was no different in my case.

---

6. Orlov, 1991, p. 304
7. Limbaugh, 2014

11. Separation of Stages in Powered Ascent   Mike Gruntman

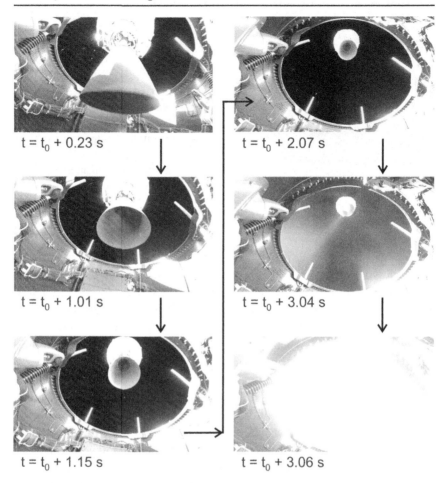

Fig. 11.6. Example of separation of the second stage during the powered ascent of Rocket Lab's *Electron* space launcher on November 20, 2020.

The separation sequence of events started at $t = t_0$ after the first stage consumed its propellant. The entire meticulously prepared process lasted a little over three seconds before the second stage ignited.

The engineering video camera on the first stage captured the sequence of events, beginning from the top-left frame. Time tags are shown immediately below the frames.

The released and pushed-out second stage first leaves the interconnecting part. After one second, it is already outside. Then, for two seconds, it drifts away from the first stage. Finally, the last two frames, which are only 20 milliseconds apart, show the ignition of the second stage, with its exhaust blinding the camera (bottom right). Video at https://youtu.be/Vpsfy4npMhY; accessed on May 11, 2021. Video credit: Rocket Lab.

Ignition of the second stage

Here are a few highlights, without details, of my second stage ignition.

As described in the Preface, in mid-March 1990, I landed at Schiphol, Amsterdam, and collected a paper ticket to Los Angeles with my name on it from a Pan American desk. The ticket was courtesy of the *Space Sciences Center*, SSC, of Darrell Judge (Fig. 11.7) from the University of Southern California in Los Angeles.

The Preface also stated that my colleagues and friends from six countries on three continents helped me in this "tunneling transition." The previous chapters mentioned some of those involved, but not all. They know who they are. I am grateful to every one of them.

A couple of days after picking up the ticket at Schiphol, I landed at LAX with $80 in my pocket. The next day, I walked into my new office: an empty desk, a chair, and a clean chalkboard (Fig. 11.8) to begin a new life from scratch.

Fig. 11.7. Director of the Space Sciences Center at USC Darrell Judge (left) and SSC's research scientist Howard Ogawa (right) in August 1990.

On the left, one can see part of an instrumentation section of a sounding rocket for measurements of the spectral distribution of the solar radiation in extreme ultraviolet.

Photograph courtesy of Mike Gruntman.

11. Separation of Stages in Powered Ascent   Mike Gruntman

Fig. 11.8. The author of this book at his desk at the Space Sciences Center (left) and on the USC campus (right) in 1991. The office and the desk did not change from his very first day in March 1990. Only the chalkboard acquired some content. Note the absence of a computer on the desk. Personal computers had not become truly "personal" yet and were often shared. Photographs from collection of Mike Gruntman.

Fig. 11.9. Tie pin (scale 2:1) commemorating an operational spacecraft, Pioneer 10, for the first time reaching a heliocentric distance of 50 AU on September 23, 1990.

Darrell Judge welcomed me to his Space Sciences Center. Actually, a scientist in his group, Howard Ogawa (Fig. 11.7), met me at LAX, as Darrell was away on a business trip. For the next three years, my office mate would be another scientist in Darrell's group, heliospheric theoretician Pradip Gangopadhyay.

The Space Sciences Center concentrated on the study of the heliosphere and the solar system interaction with the interstellar medium. It operated a Lyman-alpha photometer on the Pioneer 10 spacecraft. Working on data from this experiment became a primary task for me. Six months later, this operational spacecraft reached the distance of 50 AU from the Sun, which made me a proud participant in the celebrations of this milestone (Fig. 11.9).

The Center also had an active program of measuring spectral distributions of the solar extreme-ultraviolet radiation from sounding rockets. Don McMullin (Fig. 11.10) oversaw the laboratory with vacuum chambers and a monochromator where assembly and testing of sounding rocket payloads took place. I immediately plunged into experimental activities there, in addition, obviously, to writing proposals.

Fig. 11.10. Left: Don McMullin works on a sounding rocket payload at White Sands Missile Range in August 1990. Right: Don (left) and the author (right) at the Fourth of July celebration in 1990. From collection of Mike Gruntman.

11. Separation of Stages in Powered Ascent   Mike Gruntman

Fig. 11.11. In August 1990 SSC's Don McMullin and the author of this book drove this U-Haul truck with a sounding rocket payload section from Los Angeles to White Sands Missile Range in New Mexico. Photograph from collection of Mike Gruntman.

A great friend, Don, opened a few things for me in the new world. I cannot forget my amazement and his amusement when I discovered, with his help, that one could often open beer bottles by twisting the cap by hand. Not only had I come from continental Europe, but also its unenlightened part.

In a mere five months, Don and I drove a truck (Fig. 11.11) with a sounding rocket payload 800 miles (1300 km) from Los Angeles to Las Cruces, New Mexico, and then to *White Sands Missile Range* nearby. The U-Haul's motto painted on the front of the rented truck, "One-way anywhere," reflected the spirit of the time in my life in those days. In a month, a Black Brant rocket carried the instruments to a 200-mile altitude to measure solar EUV radiation.

Three years later in 1993, I left the Space Sciences Center and joined the USC *School of Engineering*[8] as a tenured *professor of aerospace engineering*. My primary title became *professor of astronautics* in 2004 when USC activated a new unique space engineering department, with me serving as the founding chairman. With the transition to the engineering school, the

---

8. In several years, the school would be named the *Viterbi School of Engineering*, after Andrew Viterbi.

Fig. 11.12. Working on an experiment (left) in 1994 and attending a Commencement (right) in the late 1990s. From collection of Mike Gruntman.

distance between Darrell's office and mine increased from 20 to 200 ft, but our cooperation, joint work, and friendship only strengthened.

Darrell Judge was a generous man, friend, and colleague. The USC News service released an "In memoriam" in 2014 when Darrell passed away. It quoted me:

> Professor Mike Gruntman of the USC Viterbi School of Engineering, founder of the USC Astronautics Program, recalled that many years ago, Judge helped him to escape from the former Soviet Union and make the transition to the scientific community at USC.
>
> "Darrell's generosity, hospitality and friendship have touched the lives of many people, including mine," Gruntman said. "As I started my life from scratch in the U.S., he warmly welcomed me to his group at USC and offered the hospitality of his home during my first week in Los Angeles. Darrell always encouraged me to pursue my scientific interests, which often diverged from his own.
>
> "When I found a permanent home at USC Viterbi [School of Engineering], our close scientific collaboration and friendship continued. As founding chairman of a new and unique space engineering department, I benefited immensely from Darrell's insight into administrative workings of the university."[9]

More than 30 years at USC involved me in various endeavors, including experimental work in the laboratory (Fig. 11.12, left); theoretical investigations; participation in space missions and various space-related and other

---

9. Paisley, 2014

11. Separation of Stages in Powered Ascent — Mike Gruntman

Fig. 11.13. Working with aerial reconnaissance imagery (top-right insert) on a light table at the U.S. National Archives in College Park, Md., in 2009. Photo interpreters have used such tables for decades. From collection of Mike Gruntman.

Fig. 11.14. Book signing at an AIAA conference in Orlando, Florida, in 2011. From collection of Mike Gruntman.

Fig. 11.15. The author of this book lecturing on space technology to a friendly audience in South America in 1998. From collection of Mike Gruntman.

research and development programs; consulting and advising government and industry; serving the profession; and building a new and unique (in the United States) pure space engineering academic department as the founding chairman[10] (Fig. 11.12, right). As a magazine of the engineering school observed, "It's perhaps no exaggeration to say that the department [of astronautical engineering] would not exist without Gruntman."[11]

I also worked in the national archives (Fig. 11.13) and published books on rocketry, space, missile defense, and history, including a history of one old Cold War espionage case in the 1940s involving USC (Fig. 11.14). I lectured to professional communities in government, industry, and abroad (Fig. 11.15). The latter was limited to selected friendly parts of governments in friendly countries, as I conducted my own foreign policy.

It saddens me today to observe the suicidal slide of American academia to socialist hell. "Academic freedom is withering" on campuses.[12] Many productive colleagues are usually busy with their teaching and research in a highly competitive environment while activists and radical ideologues on

---

10. Gruntman, 2007b, 2014
11. Smith, 2021, p. 50
12. e.g., Kaufmann, 2021

the left concentrate on coercive politics, taking advantage of the indifference or fear of the majority of faculty. Consequently, not that many advocate individual freedom, limits on state power, representative democracy, and a market economy.

Many universities are steadily turning into one-party enclaves. My University of Southern California is no exception. A scholarly publication shows that the Democrat-to-Republican party registration ratio at USC exceeds 25 in the combined fields of economics, history, journalism (communications), law, and psychology.[13] The causes for such a skewed ratio in a country evenly divided politically are clear. The resulting inevitable impact on the quality of scholarship and education is also obvious.

"The leftward march of the professoriate"[14] and university administrators alarms many. Hard sciences, engineering, and professional societies are becoming increasingly politicized as well. Recently, during the pandemic, the vital medical field and public health also embarked on conversion to theology, as global climate science and arms control did in the past. Replacement of quantitative test measures by subjective criteria in admissions of students and non-merit based hiring and promotion in universities, government, and industry will demolish academia and ruin businesses.

The totalitarian neo-Marxist far left drives this sinister transformation in which, as history teaches us, many protagonists will be consumed much sooner than they could imagine. The principle of karma inevitably awaits them.

Also repeating history, the defining characteristics of the present-day radical socialists include ignorance, intolerance, and lack of empathy. The more extreme they become, the more pronounced these features are. Incompetence, dysfunction, and celebration of mediocrity, which universally characterize socialist societies, grow.

A recent report of the U.S. President's Commission noted that

> Universities in the United States are often today hotbeds of anti-Americanism, libel, and censorship that combine to generate in students and in the broader culture at the very least disdain and at worst outright hatred for this country.[15]

These tragic developments are a direct indictment of the educational system and the universities in the first place. They are failing students and harming the American republic.

---

13. Langbert, et al., 2016
14. e.g., Ferguson, 2021
15. 1776 Report, 2021, p. 18

Referring to a famous appreciation of the critically important contribution of the educational system, attributed to the Duke of Wellington, George Orwell observed in 1941:

> Probably the battle of Waterloo was won on the playing-fields of Eton, but the opening battles of all subsequent wars [conducted by Great Britain] have been lost there. One of the dominant facts in English life during the past three quarters of a century has been the decay of ability in the ruling class.[16]

One does not have to be a rocket scientist to see such decay with an accompanying growth of government and slide to a socialist abyss celebrated by many in universities across the United States. It does not bode well for the free world and the world in general.

If the trend is not reversed and the free world, common sense, liberty, and human rights lose, then paraphrasing the Duke, this battle "was lost on the campuses of the Ivy League."

Sic transit gloria universitatum.

Remarkable colleagues and friends

I have met many people. Not being an operator and "politician," I openly exchanged scientific, engineering, and policy ideas and advocated and did the right things, as I saw them. Did somebody take advantage of my openness? You bet. One can find scoundrels anywhere in the world, scoundrels without borders. But the benefits outweigh the losses.

A sign on the desk of President Reagan inspired and guided me:

> There is no limit to what a man can do, or how far he can go, if he doesn't mind who gets the credit.[17]

My openness led to highly rewarding professional relations and friendships with many wonderful colleagues. It is an honor to know them.

This book mentions only some great people that I had the privilege to meet and associate with. Many remarkable highly-accomplished friends and colleagues from the second part of my life, actually the majority of them, are not mentioned at all in the story which focuses on the years at IKI.

The engine of the second stage ignited after the separation and it is still firing.

The journey and adventure continue.

Ad astra.

---

16. Orwell, 1940, p. 31
17. e.g., Wolfowitz, 2021

Appendix A

Neutral Atom Workshops in the 1980s

Scientific Programs of International Workshops and Conferences on Neutrals in Space, 1984-1989[1]

1st International Workshop
Detection of Neutral Gases in Interplanetary Space
March 26–30, 1984, Warsaw

2nd International Workshop
Problems of Physics of Neutral particles in the Solar System
September 22–27, 1985, Zakopane, Poland

3rd International Workshop
Plasma – Gas Interaction in Space
September 22–27, 1986, Guenzburg, Germany

4th International Workshop
The Interaction of Neutral Gases with Plasma in Space
September 27–October 2, 1987, Radziejowice, near Warsaw,

5th International Workshop
Space Gas Dynamics Conference
October 10 – 14, 1988, Moscow (science program not available)

1st COSPAR Colloquium
Physics of the Outer Heliosphere
September 19–22, 1989, Serock near Warsaw
(science program published in the proceedings[2])

---

1. The list of presentations at the first four neutral-atom meetings is based on the materials of the 4th workshop (Bzowski et al., 1987) and some records in the archives of the author. I could not locate the program of the 5th meeting.
2. Grzedzielski and Page, 1990

Appendix A Neutral Atom Workshops in the 1980s

Scientific Program
1st International Workshop on
Detection of Neutral Gases in Interplanetary Space
March 26–30, 1984, Warsaw

H. J. Fahr
High energy neutrals from the interface between the solar wind and the interstellar medium

V. B. Baranov
Influence of the properties of the local interstellar medium on the penetration of neutral hydrogen into the solar system

S. Bleszynski, S. Grzedzielski
Monte-Carlo calculations of the fluxes of low and high energy neutrals from the heliopause region

H. J. Fahr
Effects of elastic collisions on the distribution of interstellar neutrals in the heliosphere

H. Rosenbauer, M. Witte
Detection of neutral He atoms (GAZ-1 instrument)

M. A. Gruntman, V. B. Leonas
Measurement principle, construction and functional solutions of the GAZ-2 instrument

S. Bleszynski, S. Grzedzielski, D. Rucinski
Expected S/N ratios for neutral solar wind measurements

H. J. Fahr
Plasma-dust interaction as source of neutrals in interplanetary space

S. Grzedzielski
Possibilities of detection of heavy neutrals from the planetary magnetospheres

M. Banaszkiewicz, S. Grzedzielski
Orbits of neutrals from the Jovian planets: possibility of energy resolution

M. A. Gruntman, V. B. Leonas
   Technical possibilities offered by the interplanetary mission: feasibility of coordinated experiments

D. Rucinski
   Time-dependent effects in the distribution of neutrals caused by solar cycle activity

R. Dus
   Detection of low energy molecules by adsorption techniques

J. Ziolkowski
   Non-gravitational effects in the orbits of planetoids

W. Macek
   Planetoids as possible sources of neutral gases

M. A. Gruntman, M. Hlond, M. Rekawek
   Proposed scheme of the GAZ-3 unit

Scientific Program
2nd International Workshop on
Problems of Physics of Neutral Particles in the Solar System
September 22–27, 1985, Zakopane, Poland

H. J. Fahr
 The interaction between plasma and neutrals and the critical velocity ionization effect

K. Sauer
 Neutral gas-plasma interactions

H. J. Fahr
 LISM proton density influences on the heliospheric interface

T. Roatsch
 Nonstationary plasma – neutral gas interactions at comets

H. Rosenbauer
 New aspects of the techniques of detection of interplanetary neutral particles

M. Witte
 Abilities and measurement programs of the ULYSSES neutral particle detector

M. Hlond
 Construction and tasks performed by the GAZ-E unit

P. Blum
 Reanalysis of the Mariner 10 interplanetary glow data

D. Rucinski
 Influence of the interstellar gas parameters and solar activity on the distribution of neutrals

M. Bzowski
 Effects of time and velocity dependent radiation pressure on orbits of neutrals inside the heliosphere

H. J. Fahr
ULYSSES gas experiment and predicted data

S. Bleszynski
Secondary emission induced by neutral helium beam

M. Banaszkiewicz, S. Grzedzielski
Transcharged Jovian neutrals as the source of heavy ions in the solar wind

S. Bleszynski
Exospheric and heliospheric sources of energetic neutrals

G. Lay
INTERZODIAK experiment for the study of zodiacal dust

A. Zybura, S. Grzedzielski
Electrical charging of dust in plasma

A. Zybura
Neutrals production by sputtering of the Jovian dust ring

M. Rubel
Sputtering of silicates

Scientific Program
3rd International Workshop on
Plasma – Gas Interaction in Space
September 22–27, 1986, Guenzburg, Germany

Large Scale Properties of the Heliosphere

D. L. Judge
　The distant interplanetary glow

R. Ratkiewicz-Landowska
　The heliopause configuration under the influence of cosmic magnetic fields

R. Schwenn
　New results on solar wind plasma invariants

J. Klemens
　The momentum density flow in the solar wind at large distances

P. Blum
　Anisotropies of the interplanetary Lyman-α glow

A. Himmes
　The heliospheric interface effect in a three-dimensional treatment

H. J. Fahr
　The hydrogen and oxygen transmissivity of the heliospheric interface at variable LISM plasma conditions

W. Macek
　Reconnection at the heliopause

Ionization Mechanics

G. Haerendel
　Operation scenarios of the critical velocity effect in space

D. G. Mitchell
　Detection methods of neutral energetic atoms from the earth's ring current

P. Bochsler
Ionization states in the solar corona

J. Schmid
Ionization states of iron ions in the solar wind

E. Keppler
Neutral gas detection by means of field ionization

Interaction of Neutral and Charged particles in the Heliosphere

I. Axford
Cometary plasmas

W.-H. Ip
Plasma–gas interactions in planetary magnetospheres

E. Marsch
Pick-up and collective capture of newborn ions in the solar wind

M. Aden
Coupling and saturation times for jet ions in the heliopause boundary layer

J. Ziemkiewicz
Effects of low frequency instabilities on secondary ions in the solar wind

Sources and Distribution of Neutrals in Space

S. Grzedzielski
Signatures of the extraheliospheric neutrals

S. Bleszynski
Filtration of the LISM at the heliopause

M. Banaszkiewicz
Mercury's sodium exosphere

D. Rucinski
Influences of time-dependent processes on the interplanetary helium EUV glow and the density distribution in the inner heliosphere

H. Rosenbauer
Comments on neutrals in space

Distribution of Neutrals in Space

J. Ajello
Solar cycle study of interplanetary Ly-α variations: Results of the PIONEER-VENUS mission

M. Witte
The ULYSSES neutral-gas experiment: considerations on signal to background ratios in various environments

H. U. Nass
Neutral-ion collisions in space and relevant interaction potentials

Secondary Ions in the Solar Wind

E. C. Roelof
Simulation of ENA images of the magnetosphere and comparison with ISEE results

B. Klecker
The anomalous component of cosmic rays: pick-up ions accelerated to MeV energies

E. Moebius
Secondary ions in the solar wind: helium pick-up ions and implications for the LISM

D. Hovestadt
Interaction of ions of cometary origin with the solar wind

G. Lay
He pick-up ions and their implication for the interplanetary 304 A resonance emission

Scientific Program
4th International Workshop on
The Interaction of Neutral Gases with Plasma in Space
September 27 – October 2, 1987, Radziejowice near Warsaw

Large Scale Structure of Heliosphere: Interaction with LISM

D. L. Judge, H. S. Ogawa, Space Sciences Center, University of California,[3] Los Angeles
The observed radial variations of the interstellar glow and the expected modification due to a solar shock

P. Blum, Institut fuer Astrophysik und Extraterrestrische Forschung der Universitaet Bonn
Neutral hydrogen and helium temperatures and velocities

D. Rucinski, Space Research Centre, Warsaw
The influence of electron ionization on interstellar helium parameters

H.-U. Nass, Institut fuer Astrophysik und Extraterrestrische Forschung der Universitaet Bonn
The "elastic" heliospheric interface

M. Bzowski, Space Research Centre, Warsaw
Dependence of the LISM velocity vector on the distance from the Sun?

H. J. Fahr, Institut fuer Astrophysik und Extraterrestrische Forschung der Universitaet Bonn
Different applications of the Newtonian approximation to the problem of counterflowlng plasmas.

R. Ratkiewicz, M. Banaszkiewicz, Space Research Centre, Warsaw
An attempt to construct a 3-dimensional model of the heliopause

S. Grzedzielski, Space Research Centre, Warsaw, S. Debowski, Warsaw University
A model of the heliospheric tail

---

3. should have been University of Southern California

Appendix A   Neutral Atom Workshops in the 1980s

Opening Lecture

W. Priester, Institut fuer Astrophysik und Extraterrestrische Forschung der Universitaet Bonn
Problems of cosmic origin: Big Bang or Big Bounce?

Neutrals in Planetary and Cometary Environment

W.-H. Ip, Max-Planck-lnstitut fuer Aeronomie, Katlenburg-Lindau
Source mechanisms of magnetospheric energetic neutrals

K. Baumgaertel, lnstitut fuer Kosmosforschung, Berlin, GDR
Model calculations for the interpretation of the AMPTE experiment

W. Macek, Space Research Centre, Warsaw
Estimation of ENA fluxes in the Earth's magnetospheric tail

I. Grochulska, Space Research Centre, Warsaw
Ordinary and thermal diffusion in the terrestrial ionosphere and plasmasphere

M. Banaszkiewicz, S. Grzedzielski, Space Research Center, Warsaw
Emission of neutrals from the Io torus

V. B. Baranov, Institute for Problems in Mechanics of the USSR Academy of Sciences, Moscow
Solar wind interaction with the cometary atmosphere. Theory and comparisons with the experiments

T. Roatsch, lnstitut fuer Kosmosforschung, Berlin, GDR
Gasdynamic calculations on the solar wind interaction with an unmagnetized body

M. Staniucha, M. Banaszkiewicz, Space Research Centre, Warsaw
Disruption of the outer Oort cloud due to encounters with giant molecular clouds

Neutrals in Solar Wind Plasma

Yu. G. Malama, Institute for Problems in Mechanics of the USSR Academy of Sciences, Moscow
Statistical simulation of the interaction between interstellar hydrogen and solar wind

W.-H. Ip, Max-Planck-Institut fuer Aeronomie, Katlenburg-Lindau
Neutral hydrogen atoms in the solar corona

H. J. Fahr, Institut fuer Astrophysik und Extraterrestrische Forschung der Universitaet Bonn; J. Ziemkiewicz, Space Research Centre, Warsaw
Heating of the solar wind by the pick-up ion protons of interstellar origin

S. Grzedzielski, Space Research Centre, Warsaw
Production of ENA in distant solar wind

K. Scherer, Institut fuer Astrophysik und Extraterrestrische Forschung der Universitaet Bonn
Non-linear plasma wave interaction forms

Detection of Neutrals by In Situ Techniques

V. B. Leonas, Institute for Problems in Mechanics of the USSR Academy of Sciences, Moscow
Collisional processes related to the studies of the energetic neutral atom fluxes in the interplanetary space

H. Rosenbauer, Max-Planck-Institut fuer Aeronomie, Katlenburg-Lindau
Feasibility of measurements of very low fluxes of energetic neutrals

K. C. Hsieh, Department of Physics, University of Arizona, Tucson
Different approaches and their problems in the detection of energetic neutrals

M. A. Gruntman, V. B. Leonas, Institute for Problems in Mechanics of the USSR Academy of Sciences, Moscow
The review of opportunities to measure fluxes of energetic neutral atoms at colinear libration points L2

Appendix A   Neutral Atom Workshops in the 1980s

M. A. Gruntman, Institute for Problems in Mechanics of the USSR Academy of Sciences, Moscow
Experimental opportunity of planetary imaging in energetic neutral atoms

G. Lay, Institut fuer Astrophysik und Extraterrestrische Forschung der Universitaet Bonn
Resonance spectroscopy as a means of interplanetary gas analysis

L. Stobinski, Space Research Centre, Warsaw
Detection of atomic and molecular hydrogen in space using adsorption-desorption method

Interplanetary Dust: Dynamics and interaction with plasma

M. Rubel, Space Research Centre, Warsaw; B. Emmoth, H. Bergsaker, Research Institute of Physics, Stockholm
Sputtering of silicates in TRIM-CASC calculations

M. Rubel, Space Research Centre, Warsaw, B. Emmoth, H. Bergsaker, Research Institute of Physics, Stockholm
Ion beam irradiation of natural aluminosilicates

M. Banaszkiewicz, Space Research Centre, Warsaw; I. Kapisinski, Astronomical Institute of Slovak Academy of Sciences, Bratislava
Collisional dynamics of interplanetary dust

Appendix B

Brief History of Energetic Neutral Atoms

Excerpts (Section III, pp. 3625-3628 and references) from a review article "Energetic neutral atom imaging of space plasmas," *Review of Scientific Instruments*, v. 68, pp. 3617-3656, 1997; doi:10.1063/1.1148389, with the permission of AIP Publishing.

This appendix does not show a few figures referred to in the text; one needs to access the original article.

III. Brief history of experimental study of ENAs in space

While there are a number of publications on different aspects of the history of solar-terrestrial and magnetospheric physics [65, 66, 194-198], the story of ENA experimental study has never been told in detail. The presence of ENAs in the terrestrial environment was reliably established for the first time in 1950 by optical recording of Doppler-shifted hydrogen Balmer Hα emission (6563 A) in an aurora [199, 200]. The precipitating hydrogen ENAs are born in the charge exchange between energetic protons and neutrals of the upper atmosphere and exosphere. Balmer Hα emission is in the visible wavelength range. It can be optically detected from the ground, and auroral emissions were extensively used to study the characteristics of energetic particles [194, 201]. While hydrogen Balmer lines were observed in the auroral regions since the late 1930s [202], it was not until 1950 that a Doppler-shifted Hα line was unambiguously explained with the presence of hydrogen ENAs [199].

The importance of ENA production processes for the magnetosphere was understood [203] by noting that the proton–hydrogen atom charge exchange cross section was rather high for collision velocities less than the electron velocity in a Bohr orbit, i.e., for protons with energies <25 keV

Appendix B

Brief History of ENAs

Fig. App. B. The cover page (left), first page (center), and first page of section III "Brief history of experimental study of ENAs in space" (right) of a review article on energetic neutral atoms in space in *Review of Scientific Instruments*, 1997. Credit: American Institute of Physics.

(Fig. 2; not shown). Charge exchange determines many important properties of geomagnetic storms. The "main phase" of a geomagnetic storm may last from 12 to 24 h, and it is characterized by a weakening of the geomagnetic field. The main phase is usually followed by a "recovery phase," when a gradual field recovery toward the initial undisturbed value of the geomagnetic field is observed. The recovery time constant may be 1 day or sometimes longer.

It was suggested for the first time in 1959 that charge exchange between the magnetic storm protons and neutral atmospheric hydrogen atoms provided the mechanism for the recovery phase [77]. The charge exchange process leads to the production of fast hydrogen atoms and observation of such atoms was first proposed in 1961 as a tool to study the proton ring current present during a magnetic storm [78]. A source of ENAs beyond the magnetospheric boundary, viz., charge exchange between the solar wind and the escaping hydrogen geocorona, was also identified for the first time [78]. The concept of imaging the magnetospheric ring current in ENA fluxes from outside [13] and, "in a limited fashion," from inside [14] was introduced much later in 1984.

The presence of atomic hydrogen in interplanetary space was first derived [185] in 1963 from sounding rocket measurements [204] of Doppler-broadened hydrogen Ly-α (1216 A) radiation (see also review of the early study of extraterrestrial Ly-a radiation [205]). It was recognized since the late 1950s that Doppler shift measurements could distinguish between the telluric (geocorona) and interplanetary hydrogen [206]. The emerging concept of the heliosphere [184] was extended in 1963 by the suggestion that about half of the solar wind protons would reenter the solar cavity in the form of hydrogen ENAs (with 3/4 of the initial solar wind velocity) as a result of processes at and beyond the solar wind termination region [185]. It was established later that an interplanetary glow in the hydrogen and helium resonance lines was produced by resonant scattering of the solar radiation by interstellar gas directly entering the solar system [141-144, 158-161]. The "returning" neutral solar wind flux [185] is believed to be significantly smaller and highly anisotropic [9].

It was also suggested in the early 1960s that a large number of neutral atoms could be present in the solar wind as transients due to ejection of solar matter in violent events [207, 208]. Neutral hydrogen atoms in solar prominences are observed optically, and it was argued that they may reach 1 AU. The follow-on calculations [209, 210] showed that most of the neutral atoms would not survive travel to 1 AU because of ionization by solar EUV radiation and electron collisions. The neutral component of the solar wind is born mostly in charge exchange between the solar wind ions and interstellar gas filling the heliosphere [12, 142, 178].

Direct ENA measurements promised exceptional scientific return, but the necessary instrument development was only started in the late 1960s by Bernstein and co-workers [119, 211-213] at TRW Systems Inc., Redondo Beach, California. Direct, in situ measurement of ENA fluxes in space was first attempted on April 25, 1968 in a pioneer rocket experiment [119]. The first dedicated ENA instrument was launched on a Nike-Tomahawk sounding rocket from Fort Churchill, Manitoba, Canada. This experiment was followed by the launch of a similar instrument on a Javelin sounding rocket on March 7, 1970 to an altitude of 840 km at Wallops Island off the coast of Virginia [214]. The experiments detected hydrogen ENA fluxes in the range 10^5–10^9 cm^{-2} s^{-1} sr^{-1} with energies between 1 and 12 keV. The reported ENA fluxes were considered by many as excessively high, however these results have never been directly challenged in the literature.

The first ENA instrument [212] is shown in Fig. 8 (not shown here). It had many features and introduced many components and techniques that would later on become widely used by various modern ENA instruments. The instrument was based on foil stripping of ENAs and subsequent analysis of the resulting positive ions. Electrostatic deflection plates were used to remove incident protons and electrons with energy <25 keV at the entrance (the instrument also had an additional 100 G magnetic field to remove electrons with energies <50 keV). The deflection plates served to define the solid angle of the instrument. ENAs were stripped passing an ultrathin (2 μg/cm^2) carbon foil mounted on an 80% transparent grid. A hemispherical energy analyzer, which focused the ions in one dimension, was used for energy analysis. The stripped ENAs, protons, in a selected energy range, passed through the analyzer and were counted by a channel electron multiplier (CEM). Use of two additional identical instrument sections without deflecting voltage and without an ultrathin foil allowed the simultaneous measurement of proton fluxes and the monitoring of the background count rate during the experiment, respectively.

An attempt to measure ENAs was made in the RIEP experiment (a Russian acronym for "registerer of intensity of electrons and protons") on the Soviet Mars-3 interplanetary mission (launched May 28, 1971; entered low Mars orbit on December 2, 1971). The experiment was designed to measure the energy distribution of plasma ions and electrons in the Mars' environment as well as in the solar wind during interplanetary coast. The RIEP instrument consisted of eight separate cylindrical electrostatic energy-per-charge analyzers, each followed by a CEM to count particles [215]. Each analyzer unit was designed to measure charged particles of a selected energy-per-charge ratio. Two ultrathin carbon foils (150 A ≈ 3.5 μg/cm^2) were installed in front of two of the eight analyzers. A comparison of count rates from analyzers with and without foil while measuring particles with the same energy-per-charge ratio was expected to provide information on high-intensity neutral atom fluxes. No charged particle deflectors were used in front of the ultrathin foils, and the experiment failed to establish ENA fluxes, which are usually relatively weak.

Another ENA instrument, a slotted-disk velocity selector, was successfully built and mechanically and electrically tested in a rocket flight in 1975 [181]. This narrow (FOV) instrument, which demonstrated efficient rejection of charged particles and photons, was especially suitable for measurement of ENAs with velocities <500 km/s (E<1.3 keV/nucleon). It is interesting (see below) that measurements of the interstellar helium flux and of the neutral component in the solar wind were considered as possible applications [181]. Apparently due to large size, mass, and power consumption as well as the torque exerted on a spacecraft (the instrument included at least two disks 16.24 cm in diameter on a shaft 74 cm long spinning at 4.5×10^4 rpm) this instrument was never used for ENA measurements, and the technique development was discontinued.

The interest to the concept of a mechanically moving ENA velocity selector was recently revived by suggestion to use unconventional high-frequency mechanical shutters mounted on ceramic piezoelectric crystals [216]. Such an approach seems to be especially promising for the study of low-energy ENAs, but further development is needed to demonstrate its feasibility.

The initial ENA measurements [119, 214] were not repeated and/or independently verified, and they were largely ignored by the space community. Exceptionally strong EUV/UV radiation background was identified as a major obstacle for reliable ENA measurements in space, and experimental difficulties were perceived as insurmountable by many at that time.

Three groups accepted the experimental challenge and continued independent development of dedicated ENA instrumentation in the late 1970s. A group at the Max-Planck-Institute for Aeronomy (MPAe), Lindau, Germany targeted direct in situ detection of interstellar helium flowing into the solar system [174, 175]. (The experiment was suggested for the first time in 1972 in a proposal by H. Rosenbauer, H. Fahr, and W. Feldman.) Another group at the Space Research Institute (IKI), Moscow, USSR, planned to measure the neutral component in the solar wind and heliospheric ENAs [177, 182, 217]. It is interesting that the Moscow group initially considered also direct, *in situ* detection of the interstellar helium flux [218] but ultimately decided to concentrate on the neutral solar wind (NSW) and heliospheric ENAs. The NSW experiment was actively supported by a group at Space Research Center in Warsaw, Poland, which also theoretically studied heliospheric ENAs and ENAs produced in the giant planet magnetospheres [219-221]. A third group at the University of Arizona focused on the possibilities of measuring ENAs in geospace [222].

The GAS instrument [175] to directly detect fluxes of interstellar helium (E=30-120 eV) had a dramatic history of being completely redesigned and built within a record 3 month period in an "almost super-human effort"[176], which had become necessary to realize the original experimental concept on the European-built Ulysses probe after cancellation of the U.S. spacecraft. The GAS was then successfully flown [153, 154, 176] on Ulysses which was launched in 1990 after many delays [223], including the one caused by the Space Shuttle Challenger explosion. The neutral helium instrument is based on secondary ion emission from a specially prepared surface [175, 176]. The experiment has produced unique data on interstellar helium characteristics and ENAs emitted from Jupiter's Io torus and continues operating successfully [154].

A new ultrathin-foil based "direct-exposure" technique, which does not require stripping of incoming ENAs, was developed by the Moscow group to detect the neutral solar wind [177, 182, 217]. NSW measurements can be performed from an interplanetary or high-apogee earth-orbiting spacecraft by pointing the instrument several degrees off the sun. The possibility of taking advantage of the aberration caused by the earth's motion around the sun was first suggested in 1975 [181] and was independently "rediscovered" later [177, 182].

The NSW instrument was built for the Soviet Relikt-2 mission which was originally planned to be launched in 1987 [182]. The initial experiment also included measurements of ENAs from the heliospheric interface and ENAs emitted from the terrestrial magnetosphere. Detection of ENAs escaping Jupiter and Saturn was also expected [182, 220, 224]. The Relikt-2 mission was postponed many times, and is still awaiting launch (now scheduled for 1999).

The NSW instrument included a diffraction filter [182, 225] to significantly increase the ENA-to-EUV/UV photon ratio in the sensor. The filter was optional for the planned measurements [182], and it was not fully developed, when the instrument was built. A new diffraction filter technology is emerging [226, 227], and a new generation of diffraction filters for ENA instruments is currently being developed and evaluated (see Sec. VI F) [228-231]. An introduction of the diffraction filters opens the way to take full advantage of a highly efficient direct-exposure technique.

ENA instrument development in the late 1970s and early 1980s did not attract much attention in the space community, and did not enjoy enthusiastic support of the funding agencies. However, the importance of ENAs for mass, energy, and momentum transport in space was established and new opportunities offered by remote ENA imaging in separating spatial variations from temporal ones in space plasmas were gradually recognized.

Several measurements of large fluxes of ions (E<10 keV) near the equator at altitudes below 600 km had been reported since the early 1960s [232]. Fluxes of high-energy ions (0.25<E<1.5 MeV) were measured later at low altitudes by the German AZUR spacecraft [124,233], and the ions with energies down to 10 keV were detected there in 1973 [234]. Theoretical calculations predicted a short lifetime for such low altitude protons near the equator due to collisions with atmospheric particles. The proton loss thus required an injection of protons in a limited region below 600 km. The required proton source to compensate proton loss due to interaction with atmosphere must also be atmosphere dependent. The explanation, found in 1972 [124], suggested that trapped energetic ions in the ring current at much higher altitudes produce ENAs in charge exchange, and a fraction of these ENAs reaches low altitudes where they are re-ionized by charge exchange and are consequently trapped by magnetic field (Fig. 6).

In the heliosphere, the neutral solar wind is believed to provide a transport mechanism similar to that of magnetospheric ENAs [180]. As the solar wind expands toward the boundaries of the heliosphere, the neutral fraction in the solar wind gradually increases to 10%–20% at the termination shock (Fig. 7; not shown). While the solar wind plasma flow is terminated by the shock, the solar wind ENAs easily penetrate the region of the heliospheric interface and enter the LISM, sometimes called "very local" interstellar medium (VLISM). The solar wind ENAs thus interact with the approaching local interstellar plasma via charge exchange with plasma protons. Hence, the boundary of the region of the sun's influence, the solar system "frontier," extends further into "pristine" interstellar medium. The significance of this neutral solar wind effect [180] on LISM was recently confirmed by detailed computer simulations of the heliospheric interface [149, 150].

The interest in ENA characteristics and instrument development began to grow after the serendipitous discovery of ENAs made by energetic particle instruments on several spacecraft. Energetic particle instruments are usually based on solid state detectors and often capable of discriminating against electrons. However, such instruments generally cannot distinguish between a charged particle and an ENA. Therefore an energetic particle detector would efficiently serve as an ENA detector only in the absence of the normally abundant ions. Only high-energy ENAs (E>10-20 keV/nucleon) can be detected by such instruments.

Analysis [235] of inconsistencies in interpretation of energetic particle measurements by the IMP 7 and 8 satellites (at ~30-35 R_E from the Earth) led to the conclusion that a certain fraction of counts during periods of very low fluxes [236] was caused by radiation belt-produced ENAs with energies 0.3-0.5 MeV.

Both energetic ions and neutrals were detected during the magnetic storm in 1982 by the SEEP instrument on the S81-1 spacecraft at low altitude. A double charge-exchange mechanism (Fig. 6; not shown) was invoked to explain the observations, and it was also suggested that measurements of energetic neutral atoms at low altitude "might be able to image, in a limited fashion, the integral ion intensities of the ring current as a function of latitude and longitude" [14].

ENA fluxes of nonterrestrial origin were detected on a Voyager 1 spacecraft during flybys of Jupiter [237] and Saturn [238] in March 1979 and November 1980, respectively. The low energy charged particle (LECP) instrument on Voyager-1 included a silicon detector to accumulate counts from eight separate directions in the ecliptic plane [239]. The detector was designed to measure ions (electrons were swept away by magnetic field) with energies >40 keV.

During the approach to Jupiter, when Voyager 1 was still outside the gigantic Jovian magnetosphere, an excess count rate was measured in the sector containing the planet in its FOV. The detector characteristics limit the possible sources of the excess counts to energetic ions, x rays, and ENAs. No energetic ion fluxes with required intensity and energy were expected at the location of measurements since there were no magnetic field lines connecting to Jupiter [237]. It is known that x-rays can be generated in planetary magnetospheres by precipitating energetic electrons, usually in polar regions. Consideration of x-ray generation showed that the required fluxes of electrons were several orders of magnitude higher than those found in the Jovian magnetosphere. The conclusion of data analysis was that "the only remaining possibility of explaining the excess counts ... is energetic neutral atoms" [237]. The energy dependence of the observed ENA spectra was similar to the one established for energetic ions in the magnetosphere during the close flyby.

A similar excess count rate was measured one and a half years later during the flyby of Saturn [238]. Here again the observed count rate, if interpreted in terms of x rays, cannot be reasonably related to precipitating magnetospheric electrons. It was concluded that "charge exchange of energetic ions with satellite tori is an important loss mechanism at Saturn as well as at Jupiter"[238].

Possible ENA signatures in the experimental data obtained by energetic particle instruments [240] on IMP 7,8 and ISEE 1 were analyzed in 1982 [241, 242]. Detection of ENAs (E~50 keV) was unambiguously established, and their source was identified as the ring current in the terrestrial magnetosphere [107]. Coarse spatial information on ENA producing regions was derived and ENA energy distribution and mass composition were determined. The ENA measurements were made from positions where magnetic field lines allowed only negligible fluxes of energetic ions, so that the detectors counted only ENAs. The analysis [107] of these measurements was a major milestone in validating the idea that the global magnetospheric processes can be efficiently studied remotely by measuring ENAs. It was also suggested that ENA imaging could be used to study the magnetospheres of Jupiter and Saturn [107].

The follow-on analysis of the ISEE 1 data demonstrated the powerful potential of the ENA detection as an imaging technique by reconstructing the first ENA global image of the storm-time ring current (at E~50 keV) [79]. The instrument [240] on ISEE 1 was capable of measuring the incoming ENA flux, and imaging was performed by a combination of spacecraft spin and instrument axis scanning by a moving motorized platform. A procedure for computer simulation of all-sky ENA images was established and the theoretically predicted images were compared with the ring current image obtained by ISEE 1 at a radial distance of 2.6 R_E during a 5 min observation.

The analysis of excessive count rates [237, 238] detected by Voyager 1 during Jupiter and Saturn flybys allowed one to determine some ENA characteristics in the vicinity of the giant planets [108]. Further computer simulations of expected ENA emissions demonstrated the efficiency of ENA imaging as a tool to study the magnetospheres of Jupiter and Saturn from flyby spacecraft and orbiters [109, 110, 224, 243].

Use of energetic particle instruments for detection of high-energy ENA fluxes in the absence of ions became an established experimental technique. The storm-time ENA images of the polar cap [244] were recently obtained by the CEPPAD experiment [245] on the POLAR spacecraft. The instrument cannot distinguish between ions and neutrals, and the ENA images were recorded during the portions of the spacecraft orbit where the fluxes of the charged particles were very low.

An instrument with a dedicated high-energy ENA (20<E<1500 keV) detection channel was flown on the CRRES satellite in 1991 [21, 125]. The neutral channel that consisted of a magnetic deflector followed by a solid-state detector measured ENA fluxes precipitating to low altitudes at the equatorial regions. A small geometrical factor put limit to the imaging capabilities of the instrument. A conceptually similar ENA instrument was recently flown on ASTRID [22, 23]. A more sophisticated high energy particle (HEP) instrument (10-100 keV ENAs) was launched on the GEOTAIL spacecraft in 1992 [20].

The current phase in the study of ENAs in space plasmas is characterized by extensive computer simulations of ENA images, novel instrument development, and the preparation and planning of a number of dedicated ENA space experiments. The concept of global ENA imaging of the magnetosphere is firmly established [1-5, 100, 111, 112]. Computer simulations of low-energy [113-115] and high-energy [111, 116, 118, 246, 247] ENA imaging emphasized the importance of ENA measurements throughout a wide energy range from few eV up to several hundred keV. Various aspects

Appendix B   Brief History of ENAs

of heliospheric ENA global imaging were theoretically studied for various species and energy ranges [6-12, 157, 192, 248].

The first ENA detections obtained from planetary magnetospheres in the early 1980s triggered new interest in developing dedicated ENA instrumentation: various techniques for detection and imaging of low- and high-energy ENAs were proposed and their laboratory evaluation was begun [243, 249-254]. Several new research groups have entered the ENA-imaging field since then and significant improvements of the known experimental techniques as well as a number of innovative approaches were proposed, especially for low-energy ENAs [4, 10, 125, 216, 225, 228-230, 254-258].

NASA recently selected an IMAGE mission to perform comprehensive imaging of the terrestrial magnetosphere in EUV, FUV, and ENA fluxes. A sophisticated first large size ENA camera INCA [24] on Cassini will soon perform ENA imaging of the Saturnian magnetosphere and the exosphere of the Saturnian moon Titan [259].

References

1. E. C. Roelof and D. J. Williams, Johns Hopkins APL Tech. Dig. 9, 144 (1988).
2. E. C. Roelof and D. J. Williams, Johns Hopkins APL Tech. Dig. 11, 72 (1990).
3. J. W. Freeman, Jr., in *Physics and Astrophysics from a Lunar Base*, AIP Conference Proceedings No. 202, edited by A. E. Potter and T. L. Wilson, (AIP, New York, 1990), pp. 9-16.
4. D. J. McComas et al., Proc. Natl. Acad. Sci. USA 88, 9598 (1991).
5. D. J. Williams et al., Rev. Geophys. 30, 183 (1992).
6. K. C. Hsieh et al., Astrophys. J. 393, 756 (1992).
7. Hsieh, K. C. et al., in *Solar Wind Seven*, edited by E. Marsch and R. Schwenn (Pergamon, New York, 1992), pp. 365-368.
8. E. C. Roelof, in *Solar Wind Seven*, edited by E. Marsch and R. Schwenn (Pergamon, New York, 1992), pp. 385-394.
9. M. A. Gruntman, Planet. Space Sci. 40, 439 (1992).
10. M. A. Gruntman, Planet. Space Sci. 41, 307 (1993).
11. K. C. Hsieh and M. A. Gruntman, Adv. Space Res. 13 (6), 131 (1993).
12. M. A. Gruntman, J. Geophys. Res. 99, 19213 (1994).
13. E. C. Roelof, Eos Trans. AGU 65, 1055 (1984).
14. H. D. Voss et al., Planetary Plasma Environments: A Comparative Study, Proceedings of Yosemite '84 edited by C. R. Clauer and J. H. Waite, Jr. (AGU, Washington, DC, 1984), pp. 95-6.

20. T. Doke et al., J. Geomag. Geoelectr. 46, 713 (1994).
21. H. D. Voss et al., J. Spacecr. Rockets 29, 566 (1992).
22. S. Barabash et al., Abstracts, Chapman Conference on Measurement Techniques for Space Plasma, (AGU, Santa Fe, NM, 1995), p. 50.
23. O. Norberg et al., Proceedings of the 12th European Symposium on Rocket and Balloon Programmes and Related Research, Lillenhamer, Norway, 1995, pp. 273-7.

24. D. G. Mitchell et al., Opt. Eng. 32, 3096 (1993).

65. D. Stern, Rev. Geophys. 27, 103 (1989).
66. D. Stern, Rev. Geophys. 34, 1 (1996).

77. A. J. Dessler and E. N. Parker, J. Geophys. Res. 64, 2239 (1959).
78. A. J. Dessler et al., J. Geophys. Res. 66, 3631 (1961).
79. E. C. Roelof, Geophys. Res. Lett. 14, 652 (1987).

100. D. J. Williams, in *Magnetospheric Physics*, edited by B. Hultqvist and C.-G. Falthammer (Plenum, New York, 1990), pp. 83-101.

107. E. C. Roelof et al., J. Geophys. Res. 90, 10991 (1985).
108. A. F. Cheng, J. Geophys. Res. 91, 4524 (1986).
109. K. C. Hsieh and C. C. Curtis, Geophys. Res. Lett. 15, 772 (1988).
110. A. F. Cheng and S. M. Krimigis, in *Solar System Plasma Physics*, edited by J. H. Waite Jr., J. L. Burch, and R. L. Moore (AGU, Washington, DC, 1989), pp. 253-260.
111. A. F. Cheng et al., Remote Sens. Rev. 8, 101 (1993).
112. B. R. Sandel et al., Remote Sens. Rev. 8, 147 (1993).
113. M. Hesse et al., Opt. Eng. 32, 3153 (1993).
114. K. R. Moore et al., Instrumentation for Magnetospheric Imagery, Proc. SPIE 1744 edited by S. Chakrabarti (SPIE, Bellingham, WA, 1992), pp. 51-61.
115. K. R. Moore et al., Opt. Eng. 33, 342 (1994).
116. C. J. Chase and E. C. Roelof, Johns Hopkins APL Tech. Dig. 16, 111 (1995).

118. C. J. Chase and E. C. Roelof, Adv. Space Res. 18 (1997).
119. W. Bernstein et al., J. Geophys. Res. 74, 3601 (1969).

124. J. Moritz, Z. Geophys. 38, 701 (1972).
125. H. D. Voss et al., Opt. Eng. 32, 3083 (1993).

141. H. J. Fahr, Space Sci. Rev. 15, 483 (1974).
142. T. E. Holzer, Rev. Geophys. Space Phys. 15, 467 (1977).
143. G. E. Thomas, Ann. Rev. Earth Planet. Sci. 6, 173 (1978).
144. J.-L. Bertaux, IAU Colloq. No. 81, NASA CP 2345, 1984, pp. 3-23.

149. V. B. Baranov and Yu. G. Malama, J. Geophys. Res. 98, 15157 (1993).
150. V. B. Baranov and Yu. G. Malama, J. Geophys. Res. 100, 14755 (1995).

153. M. Witte et al., Adv. Space Res. 13, 121 (1993).
154. M. Witte et al., Space Sci. Rev. 78, 289 (1996).

157. J. Luhmann, J. Geophys. Res. 75, 13285 (1994).
158. H. J. Fahr, Astrophys. Space Sci. 2, 474 (1968).
159. R. R. Meier, Astron. Astrophys. 55, 211 (1977).
160. F. M. Wu and D. L. Judge, Astrophys. J. 231, 594 (1979).
161. J. M. Ajello et al., Astrophys. J. 317, 964 (1987).

174. H. Rosenbauer and H. J. Fahr, "Direct measurement of the fluid parameters of the nearby interstellar gas using helium as a tracer. Experimental proposal for the Out-of-Ecliptic Mission, Internal Report" (Max-Planck-Institut fur Aeronomie, Katlenburg-Lindau, Germany, 1977).
175. H. Rosenbauer et al., ESA SP-1050, 123 (1984).
176. M. Witte et al., Astron. Astrophys. Suppl. Ser. 92, 333 (1992).
177. M. A. Gruntman and V. B. Leonas, "Neutral solar wind: possibilities of experimental investigation," Report (Preprint) 825 (Space Research Institute (IKI), Academy of Sciences, Moscow, 1983).
178. H. J. Fahr, Astrophys. Space Sci. 2, 496 (1968).

180. M. A. Gruntman, Sov. Astron. Lett. 8, 24 (1982).
181. J. H. Moore Jr. and C. B. Opal, Space Sci. Instrum. 1, 377 (1975).
182. M. A. Gruntman et al., in *Physics of the Outer Heliosphere*, edited by S. Grzedzielski and D. E. Page (Pergamon, New York, 1990), pp. 355–358.

184. W. I. Axford et al., Astrophys. J. 137, 1268 (1963).
185. T. N. L. Patterson et al., Planet. Space Sci. 11, 767 (1963).

192. A. Czechowski et al., Astron. Astrophys. 297, 892 (1995).

194. J. W. Chamberlain, Physics of the Aurora and Airglow (Academic, New York, 1961).
195. W. I. Axford, J. Geophys. Res. 99, 19199 (1994).
196. N. Fukushima, J. Geophys. Res. 99, 19133 (1994).
197. C. T. Russell, in *Introduction to Space Physics*, edited by M. G. Kivelson and C. T. Russell (Cambridge University Press, Cambridge, 1995), pp. 1-26.
198. Discovery of the Magnetosphere, edited by C. S. Gillmor and J. R. Spreiter (American Geophysical Union, Washington, DC, 1997).
199. A. B. Meinel, Phys. Rev. 80, 1096 (1950).
200. A. B. Meinel, Astrophys. J. 113, 50 (1951).
201. C. Y. Fan, Astrophys. J. 128, 420 (1958).
202. L. Vegard, Nature (London) 144, 1089 (1939).
203. G. W. Stuart, Phys. Rev. Lett. 2, 417 (1959).
204. D. C. Morton and J. D. Purcell, Planet. Space Sci. 9, 455 (1962).
205. B. A. Tinsley, Rev. Geophys. Space Phys. 9, 89 (1971).
206. F. S. Johnson and R. A. Fish, Astrophys. J. 131, 502 (1960).
207. S.-I. Akasofu, Planet. Space Sci. 12, 801 (1964).
208. S.-I. Akasofu, Planet. Space Sci. 12, 905 (1964).
209. J. C. Brandt and D. M. Hunten, Planet. Space Sci. 14, 95 (1966).
210. P. A. Cloutier, Planet. Space Sci. 14, 809 (1966).
211. R. L. Wax and W. Bernstein, Rev. Sci. Instrum. 38, 1612 (1967).
212. W. Bernstein et al., in *Small Rocket Instrumentation Techniques* (North-Holland, Amsterdam, 1969), pp. 224–231.
213. W. Bernstein et al., Nucl. Instrum. Methods 90, 325 (1970).
214. R. L. Wax et al., J. Geophys. Res. 75, 6390 (1970).
215. M. R. Ainbund et al., Cosmic Res. (English transl. of Kosmicheskie Issledovaniya) (Plenum, New York, 1973), Vol. 11, pp. 661-665
216. H. O. Funsten et al., J. Spacecr. Rockets 32, 899 (1995).

217. M. A. Gruntman and V. A. Morozov, J. Phys. E 15, 1356 (1982).

218. M. A. Gruntman, "Interstellar helium at the Earth's orbit," Report (Preprint) 543 (Space Research Institute (IKI), Academy of Sciences, Moscow, 1980).

219. S. Grzedzielski, Artificial Satellites, 18, 5 (1983).

220. S. Grzedzielski, Artificial Satellites, 19, 5 (1984).

221. S. Grzedzielski and D. Rucinski, in *Physics of the Outer Heliosphere*, edited by S. Grzedzielski and D. E. Page (Pergamon, New York, 1990), pp. 367-370.

222. K. C. Hsieh et al., "HELENA—A proposal to perform in situ investigation of NEUTRALs in geospace and the heliosphere on PPL&IPL of the OPEN mission," Proposal to NASA, 1980.

223. K.-P. Wenzel and D. Eaton, European Space Agency (ESA) Bull. 63, 10 (1990).

224. M. Banaszkiewicz and S. Grzedzielski, Ann. Geophys. (Germany) 10, 527 (1992).

225. M. A. Gruntman, EUV, X-Ray, and Gamma-Ray Instrumentation for Astronomy, Proc. SPIE 1549, edited by O. H. W. Siegmund and R. E. Rothschield (SPIE, Bellingham, 1991), pp. 385-394.

226. M. L. Schattenburg et al., Phys. Scr. 41, 13 (1990).

227. J. M. Carter et al., J. Vac. Sci. Technol. B 10, 2909 (1992).

228. E. E. Scime et al., Appl. Opt. 34, 648 (1995).

229. M. A. Gruntman, X-Ray and Extreme Ultraviolet Optics, Proc. SPIE 2515, edited by R. B. Hoover and A. B. C. Walker, Jr. SPIE, Bellingham, 1995, pp. 231-239.

230. M. A. Gruntman, Appl. Opt. 34, 5732 (1995).

231. M. A. Gruntman, Appl. Opt. 36, 2203 (1997).

232. W. J. Heikkila, J. Geophys. Res. 76, 1076 (1971).

233. D. Hovestadt et al., Phys. Rev. Lett. 28, 1340 (1972).

234. P. F. Mizera and J. B. Blake, J. Geophys. Res. 78, 1058 (1973).

235. D. Hovestadt and M. Scholer, J. Geophys. Res. 81, 5039 (1976).

236. S. M. Krimigis et al., Geophys. Res. Lett. 2, 457 (1975).

237. E. Kirsch et al., Geophys. Res. Lett. 8, 169 (1981).

238. E. Kirsch et al., Nature (London) 292, 718 (1981).

239. S. M. Krimigis et al., Space Sci. Rev. 21, 329 (1977).

240. D. J. Williams et al., IEEE Trans. Geosci. Electron. GE-16, 270 (1978).

241. E. C. Roelof and D. G. Mitchell, Eos Trans. AGU 63, 403 (1982).

242. E. C. Roelof and D. G. Mitchell, Eos Trans. AGU 63, 1078 (1982).

243. C. C. Curtis and K. C. Hsieh, in *Solar System Plasma Physics*, edited by J. H. Waite Jr., J. L. Burch, and R. L. Moore, (AGU, Washington, DC, 1989) pp. 247-251.

244. M. G. Henderson et al., Geophys. Res. Lett. 24, 1167 (1997).

245. J. B. Blake et al., Space Sci. Rev. 71, 531 (1995).

246. E. C. Roelof et al., Instrumentation for Magnetospheric Imagery, Proc. SPIE 1744, edited by S. Chakrabarti, (SPIE, Bellingham, 1992), pp. 19-30.

247. E. C. Roelof et al., Instrumentation for Magnetospheric Imagery II, Proc. SPIE 2008, edited by S. Chakrabarti (SPIE, Bellingham, 1993), pp. 202-213.

248. K. C. Hsieh et al., in *Solar Wind Seven*, edited by E. Marsch and R. Schwenn (Pergamon, New York, 1992), pp. 357-364.

249. M. A. Gruntman and V. B. Leonas, "Experimental opportunity of planetary magnetosphere imaging in energetic neutral particles," Report (Preprint) 1181 (Space Research Institute (IKI), Academy of Sciences, Moscow, 1986) (also published in Proceedings, International Workshop on Problems of Physics of Neutral Particles in the Solar System, Zakopane, 1985, pp. 151-179).

250. R. W. McEntire and D. G. Mitchell, in *Solar System Plasma Physics*, edited by J. H. Waite Jr., J. L. Burch, and R. L. Moore, (AGU, Washington, DC, 1989), pp. 69-80.

251. K. C. Hsieh and C. C. Curtis, in *Solar System Plasma Physics*, edited by J. H. Waite, Jr., J. L. Burch, and R. L. Moore (AGU, Washington, DC, 1989), pp. 159-164.

252. E. P. Keath et al., in *Solar System Plasma Physics*, edited by J. H. Waite, Jr., J. L. Burch, and R. L. Moore (AGU, Washington, DC, 1989), pp. 165-170.

253. M. A. Gruntman et al., JETP Lett. 51, 22 (1990).

254. D. J. McComas et al., Instrumentation for Magnetospheric Imagery, Proc. SPIE 1744, edited by S. Chakrabarti, (SPIE, Bellingham, 1992), pp. 40-50.

255. H. O. Funsten et al., Opt. Eng. 32, 3090 (1993).

256. H. O. Funsten et al., Opt. Eng. 33, 349 (1994).

257. A. G. Ghielmetti et al., Opt. Eng. 33, 362 (1994).

258. P. Wurz et al., Opt. Eng. 34, 2365 (1995).

259. A. Amsif et al., J. Geophys. Res. 102 (in press, 1997).

Appendix C

Acronyms and Abbreviations

AGU	—	American Geophysical Union
AIAA	—	American Institute of Aeronautics and Astronautics
AMOLF	—	FOM-Instituut voor Atoom- en Molecuulfysica, or FOM Institute of Atomic and Molecular Physics (FOM stands for supporting organization Fundamental Research on Matter)
APL	—	Applied Physics Laboratory
AXAF	—	Advanced X-ray Astrophysics Facility
AU	—	astronomical unit, 1 AU = 1.496×10^{11} m
BBC	—	British Broadcasting Corporation
BNL	—	Brookhaven National Laboratory
CAMAC	—	computer-aided measurement and control
CBK	—	Centrum Badań Kosmicznych (Space Research Center), Polish Academy of Sciences, Warsaw
CCD	—	charge-coupled device
CCP	—	Chinese Communist Party
CEM	—	channel electron multiplier
CEO	—	chief executive officer
CIA	—	Central Intelligence Agency
COCOM	—	Coordinating Committee for Multilateral Export Controls

Appendix C — Acronyms and Abbreviations

COSPAR	—	Committee on Space Research, International Council of Scientific Unions, or ICSU (now the International Science Council, or ISC)
CPSU	—	Communist Party of the Soviet Union
DLR	—	Deutsches Zentrum fuer Luft- und Raumfahrt (German Aerospace Center)
ENA	—	energetic neutral atom
ESA	—	European Space Agency
EUV	—	extreme ultraviolet
FAKI	—	Fakul'tet Aerofiziki i Kosmicheskikh Issledovanii (Faculty of Aerophysics and Space Research)
FIAN	—	Fizicheskii Institut Akademii Nauk (P.N. Lebedev Physical Institute of the USSR Academy of Science)
FOIA	—	Freedom of Information Act
FOV	—	field of view
FPFE	—	Fakul'tet Problem Fiziki i Energetiki (Faculty of Problems of Physics and Energetics)
FRG	—	Federal Republic of Germany (West Germany)
GDR	—	German Democratic Republic (East Germany)
GUKOS	—	Glavnoe Upravlenie Kosmicheskikh Sredstv (Chief Directorate of Space Assets, Ministry of Defense).
IAE	—	[Kurchatov] Institute of Atomic Energy
IBEX	—	Interstellar Boundary Explorer
ICBM	—	intercontinental ballistic missile
IKI	—	Institut Kosmicheskikh Issledovanii (Space Research Institute, USSR Academy of Sciences)
IMAGE	—	Imager for Magnetopause-to-Aurora Global Exploration
IOP	—	Institute of Physics (United Kingdom)
IPM	—	Institut Problem Mekhaniki (Institute for Problems in Mechanics, USSR Academy of Sciences)
IRBM	—	intermediate-range ballistic missile

JETP	—	Journal of Experimental and Theoretical Physics
JGR	—	Journal of Geophysical Research
JPL	—	Jet Propulsion Laboratory
LFTI	—	Leningrad Physical-Technical Institute
LISM	—	local interstellar medium
MASTIF	—	mass analysis of secondaries by time-of-flight technique
MCP	—	microchannel plate
MENA	—	Medium Energy Neutral Atom Imager
MFTI	—	Moskovskii Fiziko-Tekhnicheskii Institut (Moscow Physical-Technical Institute), also known as Fiztekh
MIPT	—	Moscow Institute of Physics and Technology
MOM	—	Ministerstvo Obshchego Mashinostroeniya (Ministry of General Machine Buidling)
MPAe	—	Max-Planck-Institute fuer Aeronomie (Max Planck Institute for Aeronomie)
MPE	—	Max-Panck-Institut fuer extraterrestrische Physik (Max Planck Institute for Extraterrestrial Physics)
NII	—	nauchno-issledovatel'skii institut (scientific research institute)
NIITV	—	Nauchno-Issledovatel'skii Institut Televideniya (Scientific Research Institute of Television)
NSW	—	neutral solar wind
NSZZ	—	Niezależny Samorządny Związek Zawodowy (in Polish, Independent Self-Governing Trade Union)
NTF	—	nuclear track filter
PI	—	principal investigator
PSD	—	position-sensitive detector
PTE	—	Pribory i Tekhnika Eksperimenta (Instruments and Exoerimental Techniques)
RFE	—	Radio Free Europe
RFE/RL	—	Radio Free Europe/Radio Liberty

Appendix C — Acronyms and Abbreviations

RL	—	Radio Liberty
SAO	—	Spetsial'naya Astrofizicheskaya Observatoriya (Special Astrophysical Observatory), USSR Academy of Science
SEM	—	secondary electron multiplier
SIMS	—	secondary ion mass-spectrometry
SNIIP	—	Soyuznyi Nauchno-Issledovatel'skii Institut Priborostroeniya (Union Scientific Research Institute of Instrument Building)
SPIE	—	The International Society for Optical Engineering
SSC	—	Space Sciences Center, University of Southern California
SSL	—	Space Sciences Laboratory, University of California, Berkeley
TM	—	technical memorandum
TOF	—	time-of-flight
TWINS	—	Two Wide-Angle Imaging Neutral-Atom Spectrometers
USAGM	—	United States Agency for Global Media
USC	—	University of Southern California
USL	—	Universita 17. listopadu (in Czech, The Seventeenth November University)
USSR	—	Union of Soviet Socialistic Republics
UV	—	ultraviolet
VEU	—	vtorichnyi elektronnyi umnozhitel' (secondary electron multiplier)
VNII	—	Vsesoyuznyi Nauchno-Issledovatel'skii Institut (All-Union Scientific Research Institute)
VOA	—	Voice of America
ZOMO	—	Zmotoryzowane Odwody Milicji Obywatelskiej (in Polish, Motorized Reserves of the Citizens' Militia)

Appendix D

Pronunciation guide

Not only Russian words are written in Cyrillic but they also exhibit irregular stress patterns. This guide provides the breakdown of words into syllables and shows stress (by capitalization of the stressed syllable) which should help in proper pronunciation. While breaking words into syllables is not unique, the differences practically do not affect pronunciation. *Most words are personal names, geographical names, and elements of organizational names.*

A

aerofiziki	—	a-e-ro-FI-zi-ki
Akademgorodok	—	a-ka-DEM-go-ro-DOK
Al'bert	—	al'-BERT
Aleksei	—	a-lek-SEY
al't-azimutal'nyi	—	AL'T-a-zi-mu-TAL'-nyi
Arkhyz	—	ar-KHYZ
Artsakh (Armenian)	—	ar-TSAKH
aspirantura	—	as-pi-ran-TU-ra
avtoreferat	—	av-to-re-fe-RAT
azimutal'nyi	—	a-z-mu-TAL'-nyi

B

Baikonur	—	bai-ko-NUR
Baranov	—	ba-RA-nov
baza	—	BA-za
Belotserkovsky	—	be-lo-tser-KOV-skii
Belyaevo	—	be-LYA-ye-vo
BESM	—	BESM
bezopasnosti	—	be-zo-PAS-no-sti
bol'shaya	—	bol'-SHA-ya
bol'shoi	—	bol'-SHOI
Boris	—	bo-RIS

C

Cheremushki	—	che-RYO-mush-ki
Chernyaev	—	cher-NYA-yev

Appendix D

Pronunciation Guide

D
Dolgopa	— dol-GO-pa
Dolgoprudnyi	— dol-go-PRUD-nyi
duga	— du-GA

E
edinaya	— ye-DI-na-ya
eksperimenta	— eks-pe-ri-MEN-ta
elektronno-schetnaya	— e-lek-TRON-no-SCHET-na-ya
elektronnyi	— e-lek-TRON-nyi
energetiki	— e-ner-GE-ti-ki
ES	— ye-ES
Etkin	— ET-kin
Evgenii	— ev-GE-nii
Evlanov	— ev-LA-nov

F
FAKI	— fa-KI
fakul'tet	— fa-kul'-TET
fizicheskikh	— fi-ZI-ches-kikh
fiziki	— FI-zi-ki
fiziko-tekhnicheskii	— FI-zi-ko-tekh-NI-ches-kii
Fiztekh	— fiz-TEKH
FPFE	— fe-pe-fe-E

G
Galeev	— ga-LE-yev
Galperin	— gal'-PE-rin
Georgii	— ge-OR-gii
glavnoe	— GLAV-no-e
Glavspetsstroi	— glav-spets-STROI
Gorbachev	— gor-ba-CHYOV
gosudarstvennoi	— go-su-DAR-stven-noi
gosudarstvennyi	— go-su-DAR-stven-nyi
GUKOS	— gu-KOS

I
i	— i
IPM	— i-pe-EM
IKI	— i-KI
industriya	— in-du-STRI-ya
institut	— in-sti-TUT
issledovanii	— is-SLE-do-va-nii

K
kafedra	— KA-fed-ra
Kalinin	— ka-LI-nin
Kaluzhskaya	— ka-LUZH-ska-ya
Kapitsa	— ka-PI-tsa
KGB	— ka-ge-BE
Khodarev	— KHO-da-rev
Khristianovich	— khri-sti-a-NO-vich
Khromov	— KHRO-mov
komitet	— ko-mi-TET
kosmicheskikh	— kos-MI-ches-kikh
Kosygin	— ko-SY-gin
Kurt	— KURT

L
Leonas	— le-O-nas
Lev	— LEV
Luzhniki	— luzh-ni-KI

M
Managadze	— ma-na-GA-dze
mashina	— ma-SHI-na
mashinostroeniya	—

	ma-shi-no-stro-YE-ni-ya	Prokhorov	— PRO-kho-rov
mekhaniki	— me-KHA-ni-ki		
MFTI	— em-fe-te-I	**R**	
MGU	— em-ge-U	Radioastronomicheskii —	
Mikhail	— mi-kha-IL		RA-di-o-as-tro-no-MI-ches-kii
ministerstvo	— mi-nis-TER-stvo	Roald	— ro-AL'D
Misha	— MI-sha		
MOM	— MOM	**S**	
Morozov	— mo-RO-zov	Sagdeev	— sag-DE-yev
moskovskii	— mos-KOV-skii	Sasha	— SA-sha
Mukhin	— MU-khin	Sergei	— ser-GEI
		Shapiro	— sha-PI-ro
N		Sheremetyevo	—
nauk	— na-UK		she-re-MET'-ye-vo
nizhnii	— NIZH-nii	sistema	— sis-TE-ma
nomenklatura	—	SNIIP	— sni-IP
	no-men-kla-TU-ra	sotsialisticheskaya	—
novye	— NO-vy-ye		so-tsi-a-li-STI-ches-ka-ya
		soyuznyi	— so-YUZ-nyi
O		sredstv	— SREDSTV
obratnyi	— ob-RAT-nyi	Stepanakert	— ste-pa-na-KERT
Obraztsov	— ob-raz-TSOV	steklyashka	— stek-LYASH-ka
Obrucheva	— OB-ru-che-va	Sumgait	— sum-ga-IT
obshchego	— OB-shche-go		
Oleg	— o-LEG	**T**	
Ostrov (town)	— OS-trov	teknika	— TEKH-ni-ka
otschet	— ot-SCHYOT	telesko	— te-le-SKOP
		televideniya	— te-le-VI-de-ni-ya
P		Teplyi Stan	— TYOP-lyi STAN
Petr	— PYOTR	tovarishch	— to-VA-rishch
Petrov	— pet-ROV	Tyuratam	— tyu-ra-TAM
po chut' chut'	— po-chut'-CHUT'		
priboru	— pri-BO-ru	**U**	
pribory	— pri-BO-ry	Umansky	— u-MAN-skii
problem	— prob-LEM	umnozhitel'	— um-no-ZHI-tel'
Profsoyuznaya	—	upravlenie	— u-prav-LE-ni-e
	prof-so-YUZ-na-ya	uspekhi	— us-PE-khi

287

V

Vaisberg	—	VAIS-berg
Valentin	—	va-len-TIN
Valerii	—	va-LE-rii
Vasyukov	—	va-syu-KOV
vedushchii	—	ve-DU-shchii
Velikhov	—	VE-li-khov
Vitalii	—	vi-TA-lii
Vitya	—	VI-tya
Vladas	—	VLA-das
Vladimir	—	vla-DI-mir
VNII	—	vni-I
Volodya	—	vo-LO-dya
vremya	—	VRE-mya
vsesoyuznyi	—	vse-so-YUZ-nyi
vtorichnyi	—	vto-RICH-nyi

Y

Yulii	—	YU-lii
Yuri	—	YU-rii
Yuzhnoe	—	YUZH-no-e

Z

Zelenchukskaya	—	ze-len-CHUK-ska-ya
Zelenyi	—	ze-LYO-nyi
Zolotukhin	—	zo-lo-TU-khin

Appendix E

Selected Bibliography

The 1776 Report, The President's Advisory 1776 Commission, January 2021.

M. R. Ainbund, V. G. Kovalenko, and B. V. Polenov, Characteristics of channel electron multipliers with a funnel at the input, *Instruments and Experimental Techniques*, v. 17, n. 4, part 2, pp. 1126-1128, 1974a.

M. R. Ainbund, V. P. Pronin, and V. M. Stozharov, Investigation of the zonal characteristics of channel electron multiplier with a funnel, *Instruments and Experimental Techniques*, v. 17, n. 4, part 2, pp. 1129-1131, 1974b.

M. R. Ainbund, L. S. Gorn, M. A. Gruntman, D. S. Zakharov, A. A. Klimashev, V. B. Leonas, I. P. Maslenkov, and B. I. Khazanov, Position-sensitive detector on the basis of microchannel plates, *Instruments and Experimental Techniques*, v. 27, n. 1, part 1, pp. 68-73, 1984.

S. I. Akasofu, A source of the energy for geomagnetic storms and auroras, *Planetary and Space Science*, v. 12, pp. 801-808, 1964; The neutral hydrogen flux in the solar plasma flow – I, *Planetary and Space Science*, v. 12, pp. 905-913, 1964.

H. O. Anger, Scintillation camera, *Review of Scientific Instruments*, v. 29, pp. 27-33, 1958; doi.org/10.1063/1.1715998.

Appeal to all scientists of the world, *Pravda*, p. 4, April 10, 1983.

V. B. Baranov and V. B. Leonas, Issledovanie gazodinamicheskikh protsessov na periferii Solnechnoi sistemy (Study of processes at the boundary of the Solar system), *Vestnik Akademii Nauk SSSR* (*Bulletin/Messenger of the USSR Academy of Sciences*), n. 8, pp. 126-128, 1989.

V. B. Baranov, *Iz XX v XXI Vek. Istoriya Odnoi Zhizni* (*From 20th to 21st Century. History of One Life*), Flinta, Moscow, 2009 (in Russian).

Yu. M. Baturin, *Sovetskaya Kosmicheskaya Initsiativa v Gosudarstvennykh Dokumentakh, 1946–1964* (*Soviet Space Initiative in State Documents, 1946–1964*), RTSoft, Moscow, 2008 (in Russian).

M. Begin, *White Nights* (first published in 1957), Harper & Row, New York, 1977.

E. J. Bender, Present image-intensifier tube structures, in L. M. Biberman (editor),

Appendix E Selected Bibliography

Electro-Optical Imaging: System Performance and Modeling, pp. (5-1)-(5-96), SPIE Press, Bellingham, Wash., 2000.

B. Berkowitz, *The National Reconnaissance Office at 50 Years: A Brief History*, Center for the Study of National Reconnaissance, Chantilly, VA, 2011.

L. Bittman, *The KGB and Soviet Disinformation: An Insider's View*, Pergamon, 1985.

F. I. Blatov, B. M. Glukhovskoi, M. Gruntman, A. A. Kozochkina, and V. B. Leonas, Three large-format microchannel plate stack for position-sensitive detector, *Instruments and Experimental Techniques*, v. 33, n. 3, pt. 2, pp. 647-651, 1990.

B. N. Bragin, M. M. Butslov, L. N. Ivanova, V. S. Malysheva, D. K. Sattarov, B. M. Stepanov, and D. A. El'kinson, Some characteristics of microchannel plates, *Instruments and Experimental Techniques*, v. 18, n. 1, pt. 2, pp. 223-225, 1975.

M. Bzowski, J. Ziemkiewicz, and T. Zarnowiecki, *Proceedings of the 4th International Workshop on Interaction of Neutral Gases with Plasma in Space*, Radziejowice, Poland, September 27 – October 2, 1987, Space Research Center, Polish Academy of Sciences, Warsaw, 1987.

A. S. Chernyaev, *Sovmestnyi Iskhod: Dnevnik Dvukh Epokh. 1972-1991 Gody (Joint Outcome: Diary of Two Eras. 1972-1991 years)*, Rossiyskaya Politicheskaya Entsiklopediya, 2008 (in Russian).

CIA, *Foreign Radiobroadcasting Reception Potential in the USSR*, Provisional Intelligence Report, CIA/RR PR-82, October 21, 1954.

CIA, *Ostrov MRBM Launch Complex. USSR, Launch Area No. 3*, Photographic Interpretation Report, NPIC, December 1965.

CIA, *Ostrov MRBM Launch Facility, USSR*, Photographic Interpretation Memorandum, TCS-20104/69, NPIC, April, 1969.

CIA, *Activity and Developments at Selected Soviet Research and Institutes*, RCA-09/003/xx, NPIC, February 1, 1980.

M. R. Collier, T. E. Moore, K. W. Ogilvie, D. Chornay, J. W. Keller, S. Boardsen, J. Burch, B. El Marji, M.-C. Fok, S. A. Fuselier, A. G. Ghielmetti, B. L. Giles, D. C. Hamilton, B. L. Peko, M. Quinn, E. C. Roelof, T. M. Stephen, G. R. Wilson, and P. Wurz, Observations of neutral atoms from the solar wind, *Journal of Geophysical Research*, v. 106, pp. 24893-24906, 2001; doi: 10.1029/2000JA000382.

W. B. Colson, J. McPherson, and F. T. King, High-gain imaging electron multiplier, *Review of Scientific Instruments*, v. 44, pp. 1694-1696, 1973.

Comptroller General of the United States, *U.S. Government Monies Provided to Radio Free Europe and Radio Liberty*, B-173239, Report to the Committee on Foreign Relations, United States Senate, The Comptroller General of the United States, 1972.

Constitution (basic law) of the Union of Soviet Socialist Republics, October 7, 1977 (in Russian); http://www.hist.msu.ru/ER/Etext/cnst1977.htm; accessed on September 8, 2021.

S. Courtois, N. Werth, J.-L. Panné, A. Paczkowski, K. Bartošek, J.-L. and Margolin, *The Black Book of Communism: Crimes, Terror, Repression*, Harvard Univ. Press, Cambridge, MA, 1999.

L. Cowan, X Ray/EUV optics highlighted at SPIE's annual meeting, [Optical Engineering] *OE Reports*, SPIE, No. 141, pp. 1, 6, 1995.

J. Critchlow, *Radio Hole-in-the-Head. Radio Liberty: An Insiders's Story of Cold War Broadcasting*, American University Press, Washington, D.C., 1995.

A. J. Dessler, W. B. Hanson, and E. Parker, Formation of the geomagnetic storm main-phase ring current, *Journal of Geophysical Research*, v. 66, pp. 3631-3637, 1961; doi: 10.1029/JZ066i011p03631.

A. W. Dulles, *The Craft of Intelligence*, The Lyons Press, Guilford, Conn., 2006.

H. P. Eubank, Determination of plasma ion temperatures by analysis of charge-exchange neutrals, pp. 7-16, in *Diagnostics for Fusion Experiments*, eds. E. Sindoni and C. Wharton, Pergamon Press, New York, 1979.

N. Ferguson, I'm helping to start a new college because higher ed is broken, *Bloomberg Opinion*, November 8, 2021; https://www.bloomberg.com/opinion/articles/2021-11-08/niall-ferguson-america-s-woke-universities-need-to-be-replaced, accessed on November 20, 2021.

C. Firmani, E. Ruiz, C. W. Carlson, M. Lampton, and F. Paresce, High-resolution imaging with a two-dimensional resistive anode photon counter, *Review of Scientific Instruments*, v. 53, pp. 570-574, 1982; https://doi.org/10.1063/1.1137025.

J. Frolik, *Hearing Before the Subcommittee to Investigate the Administration of the Internal Security Act and Other Internal Security Laws*, Committee on the Judiciary, United States Senate, Ninety-Fourth Congress, First Session, November 18, 1975.

J. Frolik, *The Frolik Defection. The Memoirs of an Intelligence Agent*, Corgi Books, London, 1976.

H. O. Funsten, F. Allegrini, G. B. Crew, R. DeMajistre, P. C. Frisch, S. A. Fuselier, M. Gruntman, P. Janzen, D. J. McComas, E. Moebius, B. Randol, D. B. Reisenfeld, E. C. Roelof, N. A. Schwadron, Structures and spectral variations of the outer heliosphere in IBEX energetic neutral atom maps, *Science*, v. 326, pp. 964-966, 2009; doi: 10.1126/science.1180927.

H. O. Funsten, R. W. Harper, and D. J. McComas, Absolute detection efficiency of space-based ion mass spectrometers and neutral atom imagers, *Review of Scientific Instruments*, v. 76, 053301, 2005; doi: 10.1063/1.1889465.

A. A. Galeev and G. M. Tamkovich (eds.), *Space Research Institute of the Russian Academy of Sciences. The 35th anniversary*, Space Research Institute IKI, Moscow, 1999 (in Russian and English).

G. Gamow, *My World Line. An Informal Autobiography*, Viking Press, New York, 1970.

Ya. Golovanov, *Zametki Vashego Sovremennika* (*Notes of Your Contemporary*); v. 1, 1953-1970; v. 2, 1970-1983; v. 3, 1983-2000, Dobroe Slovo, Moscow, 2001 (in Russian).

G. W. Goodrich and W. C. Wiley, Electron Multiplier, U.S. Patent 3,128,408, filed April 20, 1960 (granted April 7, 1964).

G. W. Goodrich and W. C. Wiley, Continuous channel electron multiplier, *Review of Scientific Instruments*, v. 33, pp. 761-762, 1962; https://doi.org/10.1063/1.1717958.

Yu. V. Gott, *Vzaimodeistvie Chastits s Veshchestvom v Plazmennykh Issledovaniyakh*, (*Interaction of Particles with Matter in Plasma Research*), Atomizdat, Moscow, 1978 (in Russian).

L. Grizzard, *The wit and wisdom of Lewis Grizzard*, Longstreet Press, Marietta, Ga., 1995.

T. Groseclose, *Left Turn: How Liberal Media Bias Distorts the American Mind*,

Appendix E   Selected Bibliography

St. Martin's Griffin, New York, 2011.

M. A Gruntman and A. P. Kalinin, Detection of neutral particles with energies 0.6-2.0 keV by funneled channel electron multiplier, Preprint 311 (Report #311), 23 pages, Space Research Institute (IKI), USSR Academy of Sciences, Moscow, 1977a (in Russian).

M. A. Gruntman and A. P. Kalinin, Characteristics of VEU-6 channel electron multiplier in detection of neutral particles, *Instruments and Experimental Techniques*, v. 20, n. 2, pp. 528-530, 1977b.

M. A Gruntman and A. P. Kalinin, Registration characteristics of neutral particles with power 0.6-2.0 keV channel electron multiplier with funnel, NASA Technical Memorandum TM-75540, unclas[sified], 1978.

M. A. Gruntman and A. P. Kalinin, Microchannel plate detector for ions with the energy of several keV, Preprint-496 (Report # 496), 20 pages, Space Research Institute (IKI) of the USSR Academy of Sciences, Moscow, 1979 (in Russian).

M. Gruntman, Influence of position-sensitive detector discrete structure on accuracy of coordinates determination, *Journal of Physics E*, v. 13, pp. 388-390, 1980a; doi: 10.1088/0022-3735/13/4/007.

M. A. Gruntman, The neutral component of the solar wind at Earth's orbit, *Cosmic Research*, v. 18, n. 4, pp. 649-651, 1980b.

M. A. Gruntman and A. P. Kalinin, Characteristics of a microchannel plate unit in recording heavy particles, *Instruments and Experimental Techniques*, v. 23, n. 4, part 2, pp. 991-993, 1980.

M. A. Gruntman and V. A. Morozov, Study of performance of fast H atom thin foil based energy analyzer, Preprint 667 (Report #667), 65 pages, Space Research Institute (IKI), USSR Academy of Sciences, Moscow, 1981 (in Russian).

M. A. Gruntman and V. A. Morozov, H atom detection and energy analysis by use of thin foils and TOF technique, *Journal of Physics E*, v. 15, n. 12, pp. 1356-1358, 1982; doi: 10.1088/0022-3735/15/12/021.

M. A. Gruntman, Effect of neutral component of solar wind on the interaction of the Solar system with the interstellar gas flow, *Soviet Astronomy Letters*, v. 8, n. 1, pp. 24-26, 1982.

M. A. Gruntman, Identification of ions by their masses in multichannel energy analyzers, *Instruments and Experimental Techniques*, v. 26, n. 4, pt. 2, pp. 943-945, 1983.

M. Gruntman and V.B. Leonas, Neutral solar wind. Possibility of experimental study, Preprint 825 (Report # 825), 42 pages, Space Research Institute (IKI), USSR Academy of Sciences, Moscow, 1983 (in Russian).

M. A. Gruntman, Position-sensitive detectors based on microchannel plates (review), *Instruments and Experimental Techniques*, v. 27, n. 1, pt. 1, pp. 1-19, 1984.

M. A. Gruntman, The single-electron pulse height resolution as a characteristic of open-type secondary electron multipliers, *Instruments and Experimental Techniques*, v. 28, n. 1, pt. 2, pp. 157-158, 1985.

M. Gruntman and V. B. Leonas, Experimental opportunity of planetary magnetosphere imaging in energetic neutral atoms, Preprint 1181 (Report #1181), 30 pages, Space Research Institute (IKI), USSR Academy of Sciences, 1986.

M. A. Gruntman, MASTIF: Mass analysis of secondaries by time-of-flight technique.

New approach to secondary ion mass spectrometry, *Review of Scientific Instruments*, v. 60, pp. 3188-3196, 1989; doi:10.1063/1.1140550.

M.A. Gruntman, A.A. Kozochkina, and V.B. Leonas, Statistics of secondary electron emission from thin foils, *Instruments and Experimental Techniques*, v. 32, n. 3, pp. 668-671, 1989.

M.A. Gruntman, Direct interstellar atom and heliospheric-interface ENA detection on Interstellar Probe – mission to the Solar system frontier, Report # 101, Space Sciences Center, University of Southern California, Los Angeles, October 1990.

M.A. Gruntman and M.A. Mkrtchyan, Promise of use of PSD in a telescope for sky survey in far ultraviolet, Preprint 432 (Report #432), Institute for Problems in Mechanics (IPM), USSR Academy of Sciences, Moscow, 1990 (in Russian).

M.A. Gruntman, A.A. Kozochkina, and V.B. Leonas, Multielectron secondary emission from thin foils bombarded by accelerated beams of atoms, *JETP Letters*, v. 51, n. 1, pp. 22-25, 1990a.

M.A. Gruntman, I.I. Katsan, A.A. Kozochkina, S.N. Korneev, V.B. Leonas, and A.A. Petrov, Large format wedge-and-strip collector for position-sensitive detector, *Instruments and Experimental Techniques*, v. 33, n. 4, pp. 901-904, 1990b.

M.A. Gruntman, S. Grzedzielski, and V.B. Leonas, Neutral solar wind experiment, in *Physics of the Outer Heliosphere*, Proceedings of the 1st COSPAR Colloquium, ed. by S. Grzedzielski and D.E. Page, pp. 355-358, Pergamon Press, 1990c; doi: 10.1016/B978-0-08-040780-7.50059-2.

M. Gruntman, In situ measuring of the composition (hydrogen, deuterium, and oxygen atoms) of interstellar gas, Report # 102, Space Sciences Center, University of Southern California, Los Angeles, February 1991a; updated version Report #102M, February-April, 1991a.

M. Gruntman, Submicron structures: promising filters in EUV – a review, in *EUV, X-Ray, and Gamma-Ray Instrumentation for Astronomy*, eds. O.H. Siegmund and R.E. Rothschield, Proceedings of SPIE 1549, pp. 385-394, 1991b; doi: 10.1117/12.48355.

M. Gruntman, D.L. Judge, C.C. Curtis, K.C. Hsieh, D. Shemansky, M. Seidl, A. Chutjian, and O.J. Orient, A new approach to an in situ study of neutral atomic fluxes in the solar system, *Eos Trans. AGU*, v. 72, No. 44, Fall Meeting Suppl., SH52A-10, p. 393, 1991.

M. Gruntman, Anisotropy of the energetic neutral atom flux in the heliosphere, *Planetary and Space Science*, v. 40, pp. 439-445, 1992a; doi:10.1016/0032-0633(92)90162-H.

M.A. Gruntman, A new way to measure the composition of the interstellar gas surrounding the heliosphere, Book of Abstracts, World Space Congress, Observations of the Outer Heliosphere, D.2-S.4.02, p. 427, Washington, DC, 1992b.

M.A. Gruntman, A new technique for in situ measurement of the composition of interstellar gas in the heliosphere, *Planetary and Space Science*, v. 41, n. 4, pp. 307-319, 1993a; doi: 10.1016/0032-0633(93)90026-X.

M. Gruntman, A new way to measure the composition of the interstellar gas surrounding the heliosphere, *Advances in Space Research*, v. 13, n. 6, pp. 141-144, 1993b; doi:10.1016/0273-1177(93)90403-X.

M. Gruntman, Coded-aperture technique for magnetosphere imaging: advantages and limitations, in *Instrumentation for Magnetospheric Imagery II*, ed. S. Chakrabarti, Proceedings of SPIE 2008, pp. 58-73, 1993c; doi: 10.1117/12.147644.

M. Gruntman, Transmission grating filtering and polarization characteristics in EUV, in *X-ray and Extreme Ultraviolet Optics*, ed. R.B. Hoover, A.B.C. Walker Jr., Proceedings of SPIE 2515, pp. 231-239, 1995a; doi: 10.1117/12.212592.

M. Gruntman, EUV radiation filtering by free-standing transmission gratings, *Applied Optics*, v. 34, n. 25, pp. 5732-5737, 1995b; doi: 10.1364/AO.34.005732.

M. Gruntman, Imaging of space plasmas in energetic neutral atom fluxes, *Review of Scientific Instruments*, v. 68, pp. 3617-3656, 1997a; doi:10.1063/1.1148389.

M. Gruntman, Transmission grating filtering of 52-140 nm radiation, *Applied Optics*, v. 36, n. 10, pp. 2203-2205, 1997b; doi:10.1364/AO.36.002203.

M. Gruntman, E.C. Roelof, D.G. Mitchell, H.J. Fahr, H.O. Funsten, and D.J. McComas, Energetic neutral atom imaging of the heliospheric boundary region, *Journal of Geophysical Research*, v. 106, pp. 15767-15781, 2001; doi:10.1029/2000JA000328.

M. Gruntman, *Blazing the Trail. The Early History of Spacecraft and Rocketry*, AIAA, Reston, Va., 2004.

M. Gruntman, *From Astronautics to Cosmonautics*, Booksurge, North Charleston, S.C., 2007a; ISBN 9781419670855.

M. Gruntman, The time for academic departments in astronautical engineering, Space–2007, Long Beach, Calif., AIAA–2007–6042, 2007b; doi: 10.2514/6.2007-6042.

M. Gruntman, Socks for the first cosmonaut of planet Earth, *Quest*, v. 18, n. 1, p. 44-47, 2011; also at http://astronauticsnow.com/astrosocks/gruntman_quest_v_18_n_1_2011.pdf; accessed on May 24, 2021.

M. Gruntman, Advanced degrees in astronautical engineering for the space industry, *Acta Astronautica*, v. 103 pp. 92–105, 2014; doi.org/10.1016/j.actaastro.2014.06.016.

M. Gruntman, *Intercept 1961. The Birth of Soviet Missile Defense*, AIAA, Reston, Va., 2015a.

M. Gruntman, Interstellar hydrogen ionization in the heliosheath, *Journal of Geophysical Research*, v. 120, pp. 6119-6132, 2015b; doi:10.1002/2015JA021539.

M. Gruntman, Master of science degree in astronautical engineering through distance learning, 69th International Astronautical Congress, Bremen, Germany, IAC-18-E1-4-11, 2018; also http://astronauticsnow.com/2018aste.pdf; accessed on May 1, 2021.

M. Gruntman, From Tyuratam Missile Range to Baikonur Cosmodrome, *Acta Astronautica*, v. 155, pp. 350-366, 2019; doi: 10.1016/j.actaastro.2018.12.021.

S. Grzedzielski and D.E. Page, editors, *Physics of the Outer Heliosphere*, Proceedings of the 1st COSPAR Colloquium, Pergamon Press, New York, 1990.

J. Heerikhuisen, N.V. Pogorelov, G.P. Zank, G.B. Crew, P.C. Frisch, H.O. Funsten, P.H. Janzen, D.J. McComas, D.B. Reisenfeld, and N.A. Schwadron, Pick-up ions in the outer heliosheath: a possible mechanism for the Interstellar Boundary EXplorer ribbon, *Astrophysical Journal Letters*, v. 708, L126–L130, 2010; doi:10.1088/2041-8205/708/2/L126.

D. Henninger, The Pelosi commission fiasco, *Wall Street Journal*, v. CCLXXVII, No. 123, p. A19, May 27, 2021.

P. Hertz and W. Wagner, *IBEX Proposal Debriefing Notes. Science Debrief*. Dr. Paul Hertz and Dr. Bill Wagner, NASA HQ, November 20, 2003.

K. C. Hsieh, HELENA—A proposal to perform in situ investigation of NEUTRALs in geospace and the heliosphere on PPL&IPL of the OPEN mission, Proposal to NASA, 1980.

R. V. Jones, *Most Secret War*, Coronet Books, Hodder and Stoughton, London, 1978.

O. Kalugin, *Spymaster. My Thirty-two Years in Intelligence and Espionage Against the West*, Basic Books, New York, 2009.

N.V. Karlov, L.P. Skorovarova, and N.F. Simonova (compilers), *Ya – Fiztekh (I Am Fiztekh)*, TsentrKom, Moscow, 1996 (in Russian).

N. V. Karlov and N. N. Kudryavtsev, *K istorii elitarnogo inzhenernogo obrazovaniya. (Moskovskii fiziko-technicheskii institut)* [*On the history of elite engineering education (Moscow Physical-Technical Institute)*], Preprint No. 2, MFTI-2000-2, Moscow Physical-Technical Institute, 2000 (in Russian).

N.V. Karlov, *Shershavym Yazykom Prikaza. Fiztekh. Arkhivnye Dokumenty 1938-1952 (By Rough Language of Ordinances. Fiztekh. Archival Documents from 1938-1952)*, Preprint No. 1, Moscow Physical-Technical Institute, Moscow, 2006 (in Russian).

V. L. Katayev, *A Memoir of the Missile Age. One Man's Journey*, Hoover Institution Press, Stanford, Calif., 2015.

M. Kaufman, Polish police break up a protest marking 1970 Gdansk food riots, *The New York Times*, v. CXXXIV, n. 46261, pp. A1, A5, December 17, 1984.

E. Kaufmann, Academic freedom is withering, *Wall Street Journal*, v. CCLXXVII, No. 48, p. A17, March 1, 2021.

P. Kengor, *Dupes: How America's Adversaries Have Manipulated Progressives for a Century*, ISI Books, Wilmington, Del., 2010.

S. A. Khristianovich, "Bezobrazniki" lomayut traditsii ("Scamps" break traditions), in N. V. Karlov, L. P. Skorovarova, and N. F. Simonova (compilers), *Ya – Fiztekh (I am Fiztekh)*, pp. 18-23, TsentrKom, Moscow, 1996 (in Russian).

G. F. Kennan, *Memoirs 1925-1950*, Pantheon Books, New York, 1983.

M.A. Klochko, *Soviet Scientist in Red China*, Frederick A. Praeger, New York, 1964.

S.N. Konyukhov (editor), *Prizvany Vremenem. Ot Protivostoyaniya k Mezhdunarodnomu Sotrudnuchestvu (Called by the Time. From Confrontation to International Cooperation)*, 2nd edition, ART-PRESS, Dnepropetrovsk, 2009 (in Russian).

W. G. Krivitsky, *In Stalin's Secret Service. An Exposé of Russia's Secret Policies by the Former Chief of the Soviet Intelligence in Western Europe*, Harper and Brothers Publishers, New York, 1939.

W. G. Krivitsky, *MI 5 Debriefing*, edited by G. Kern, Xenos Books, Riverside, Calif., 2004.

L. A. Kubetsky, Multiple amplifier, *Proceedings of the Institute of Radio Engineers*, v. 25, n. 4, pp. 421-433, 1937.

M. Lampton, The microchannel image intensifier, *Scientific American*, v. 145, n. 5, pp. 62-71, 1981.

M. Langbert, A. J. Quain, and D. B. Klein, Faculty voter registration in economics, history, journalism, law, and psychology, *Econ Journal Watch: Scholarly Comments on Academic Economics*, v. 13, No. 3, pp. 422–451, 2016.

M. Langbert, Homogeneous: the political affiliations of elite liberal arts college faculty, *Academic Questions*, v. 31, pp. 186-197, 2018; doi: 10.1007/s12129-018-9700-x.

V. I. Lenin to M. Gorky, Letter, September 15, 1919; in V. I. Lenin, *Complete Works*, 5th edition, v. 51, pp. 47-49, Political Literature Publishing House, Moscow, 1970.

B. Leskovar, Microchannel plates, *Physics Today*, v. 30, n. 11, pp. 42-49, 1977.

Ye. K. Ligachev and V. M. Chebrikov, On ceasing of jamming of broadcasts of radio stations "Voice of America," "BBC," "Radio Peking [Beijing]," and "Radio Korea," memo to the Central Committee of the CPSU, September 25, 1986 (in Russian); https://zen.yandex.ru/media/id/5cab82da84805b00aef6cbb6/pochemu-predsedatel-kgb-razreshil-v-sssr-vescanie-golosa-ameriki-5f6e9b9cdf292d1109f2e37f; accessed April 11, 2021.

R. Limbaugh, It's a 25th Anniversary Gorbasm!, November 10, 2014, https://www.rushlimbaugh.com/daily/2014/11/10/it_s_a_25th_anniversary_gorbasm/, accessed March 21, 2021.

C. Martin, P. Jelinsky, M. Lampton, R. F. Malina, and H. O. Anger, Wedge-and-strip anodes for centroid-finding position-sensitive photon and particle detectors, *Review of Scientific Instruments*, v. 52, pp. 1067- 1074, 1981; doi 10.1063/1.1136710.

D. J. McComas, P. A. Bochsler, L. A. Fisk, H. O. Funsten, J. Geiss, G. Gloeckler, M. Gruntman, D. L. Judge, S. M. Krimigis, R. P. Lin, S. Livi, D. G. Mitchell, E. Moebius, E. C. Roelof, N. A. Schwadron, M. Witte, J. Woch, P. Wurz, and T. H. Zurbuchen, Interstellar Pathfinder – A mission to the inner edge of the interstellar medium, CP679, Solar Wind Ten: Proceedings of the Tenth Solar Wind Conference, eds. M. Velli, R. Bruno, and F. Malara, AIP, pp. 834-837, 2003; doi: 10.1063/1.1618720.

D. J. McComas, F. Allegrini, J. Baldonado, B. Blake, P. C. Brandt, J. Burch, J. Clemmons, W. Crain, D. Delapp, R. DeMajistre, D. Everett, H. Fahr, L. Friesen, H. Funsten, J. Goldstein, M. Gruntman, R. Harbaugh, R. Harper, H. Henkel, C. Holmlund, G. Lay, D. Mabry, D. Mitchell, U. Nass, C. Pollock, S. Pope, M. Reno, S. Ritzau, E. Roelof, E. Scime, M. Sivjee, R. Skoug, T. S. Sotirelis, M. Thomsen, C. Urdiales, P. Valek, K. Viherkanto, S. Weidner, T. Ylikorpi, M. Young, and J. Zoennchen, The Two Wide-angle Imaging Neutral-atom Spectrometers (TWINS) NASA mission-of-opportunity, *Space Science Reviews*, v. 142, pp. 157-231, 2009a; doi: 10.1007/s11214-008-9467-4.

D. J. McComas, F. Allegrini, P. Bochsler, M. Bzowski, E. R. Christian, G. B. Crew, R. DeMajistre, H. Fahr, H. Fichtner, P. C. Frisch, H. O. Funsten, S. A. Fuselier, G. Gloeckler, M. Gruntman, J. Heerikhuisen, V. Izmodenov, P. Janzen, P. Knappenberger, S. Krimigis, H. Kucharek, M. Lee, G. Livadiotis, S. Livi, R. J. MacDowall, D. Mitchell, E. Moebius, T. Moore, N. V. Pogorelov, D. Reisenfeld, E. Roelof, L. Saul, N. A. Schwadron, P. W. Valek, R. Vanderspek, P. Wurz, and G. P. Zank, First global observations of the interstellar interaction from the Interstellar Boundary Explorer, *Science*, v. 326, pp. 959-962, 2009b; doi: 10.1126/science.1180906.

B. McIntyre, *The Spy and the Traitor*, Random House, New York, 2018.

S. S. Medley, A. J. H. Donné, R. Kaita, A. I. Kislyakov, M. P. Petrov, and A. L. Roquemore, Contemporary instrumentation and application of charge exchange neutral particle diagnostics in magnetic fusion energy experiments, *Review of Scientific Instruments*, v. 79, 011101, 2008; doi: 10.1063/1.2823259.

C. Meyer, *Facing Reality. From World Federalism to the CIA*, Harper and Row, New York, 1980.

S. Mickelson, *America's Other Voice. The Story of Radio Free Europe and Radio Liberty*, Praeger, New York, 1983.

E. Moebius, G. Gloeckler, M. Gruntman, and H. Fahr, A Local Interstellar Medium Explorer (LIME) mission to set a benchmark with in-situ measurements from 1–3 AU, *Eos Trans. AGU*, v. 79, n. 17, Spring Meet. Suppl., p. S269, April 28, 1998.

T. E. Moore, M. R. Collier, J. L. Burch, D. J. Chornay, S. A. Fuselier, A. G. Ghielmetti, B. L. Giles, D. C. Hamilton, F. A. Herrero, J. W. Keller, K. W. Ogilvie, B. L. Peko, J. M. Quinn, T. M. Stephen, G. R. Wilson, and P. Wurz, Low energy neutral atoms in the magnetosphere, *Geophysical Research Letters*, v. 28, pp. 1143-1146, 2001; doi: 10.1029/2000GL012500.

W. B. Morrow, Jr., J. Rennie, and W. Markey, Development and manufacture of the microchannle plate (MCP), AMSEL-NV-TR-0064, Center for Night Vision and Electro-Optics (CNVEO), Fort Belvoir, Va 22060, 1988.

R. Murphy, *Diplomat Among the Warriors*, Doubleday and Co., Garden City, NY, 1964.

N. I. Muskhelishvili, S. L. Sobolev, M. A. Lavrent'ev, A. O. Gel'fond, D. Yu. Panov, S. A. Khristianovich, F. P. Gantmakher, F. A. Terebin, and N. E. Kochin, Nuzhna vysshaya politekhnicheskaya shkola (Need for higher polytechnic school), letter to the editor, *Pravda*, n. 334 (7659), p. 3, December 4, 1938 (in Russian).

Obratnyi Otschet Vremeni. 40 Let Institutu Kosmicheskikh Issledovanii Rossiiskoi Akademii Nauk (*Reverse Countdown of Time. 40 Years of the Space Research Institute of the Russian Academy of Sciences*), Space Research Institute IKI, Moscow, 2006 (in Russian).

Obratnyi Otschet ... 2. 45 Let Institutu Kosmicheskikh Issledovanii Rossiiskoi Akademii Nauk (*Reverse Countdown ... 2. 45 Years of the Space Research Institute of the Russian Academy of Sciences*), Space Research Institute IKI, Moscow, 2010 (in Russian).

Obratnyi Otschet ... 3. 50 Let Institutu Kosmicheskikh Issledovanii Rossiiskoi Akademii Nauk (*Reverse Countdown ... 3. 50 Years of the Space Research Institute of the Russian Academy of Sciences*), Space Research Institute IKI, Moscow, 2015 (in Russian).

Obratnyi Otschet ... 4 (*Reverse Countdown ... 4*), Space Research Institute IKI, Moscow, 2016 (in Russian).

M. A. O'Grady, A sanction worth lifting, *Wall Street Journal*, v. CCLXXVIII, No. 57, p. A17, September 7, 2021.

Yu. Orlov, *Dangerous Thoughts. Memoirs of a Russian Life*, William Morrow and Company, New York, 1991.

G. Orwell, *The Lion and the Unicorn: Socialism and the English Genius* (first published 1940), in G. Orwell, Why I Write, pp. 11-94, Penguin Books, 2005.

I. M. Pacepa, *Red Horizons. The True Story of Nicolae and Elena Ceausescus' Crimes, Lifestyle, and Corruption,* Regnery History, Washington, DC, 1987.

I. M. Pacepa and R. J. Rychlak, *Disinformation. Former Spy Chief Reveals Secret Strategies for Undermining Freedom, Attacking Religion, and Promoting Terrorism*, WND Books, Washington, D.C., 2013.

M. Pack, Mainstream media partisanship comes to Voice of America, *Wall Street Journal*, v. CCLXXVII, No. 13, p. A13, January 16-17, 2021.

L. Paisley, In memoriam: Darrell Judge, USC News, August 29, 2014; https://news.usc.edu/67829/in-memoriam-darrell-judge-79/; accessed August 30, 2014

Appendix E — Selected Bibliography

T. N. L. Patterson, K. S. Johnson, and W. B. Hanson, The distribution of interplanetary hydrogen, *Planetary and Space Science*, v. 11, pp. 767-778, 1963.

M. P. Petrov, V. I. Afanasyev, F. V. Chernyshev, P. R. Goncharov, M. I. Mironov, and S. Ya. Petrov, 60 Years of neutral particle analysis: from early tokamaks to ITER, *European Physical Journal H*, v. 46, 5, 2021; doi: 10.1140/epjh/s13129-021-00009-6.

J. R. Pierce, Electron multiplier design, *Bell Laboratory Record*, v. 16, pp. 305-308, 1938.

R. Pleikis, *Radiotsenzura (Radio Censorship)*, Vilnius, 2002-2003, https://radiocenzura.tripod.com/text.htm; accessed April 5, 2021.

Polish police battled demonstrators led by Lech Walesa in Gdansk, *Wall Street Journal*, v. CCIV, n. 118, p. 1, December 17, 1984.

C. J. Pollock, K. Asamura, J. Baldonado, M. M. Balkey, P. Barker, J. L. Burch, E. J. Korpela, J. Cravens, G. Dirks, M.-C. Fok, H. O. Funsten, M. Grande, M. Gruntman, J. Hanley, J.-M. Jahn, M. Jenkins, M. Lampton, M. Marckwordt, D. J. McComas, T. Mukai, G. Penegor, S. Pope, S. Ritzau, M. L. Schattenburg, E. Scime, R. Skoug, W. Spurgeon, T. Stecklein, S. Storms, C. Urdiales, P. Valek, J. T. M. Van Beek, S. E. Weidner, M. Wuest, M. K. Young, and C. Zinsmeyer, Medium Energy Neutral Atom (MENA) imager for the IMAGE mission, *Space Science Reviews*, v. 91, pp. 113-154, 2000; doi: 10.1023/A:1005259324933.

V. P. Ponomarenko and A. M. Filachev, *Infrared Techniques and Electro-Optics in Russia: A History 1946-2006*, SPIE Press, Bellingham, WA, 2007.

P. Pressel, *Meeting the Challenge. The Hexagon KH-9 Reconnaissance Satellite*, AIAA, Reston, Va., 2013.

The promotion of peace (editorial), *Nature*, v. 142, n. 3597, October 8, 1938.

R. Reagan, Address to the Nation on Defense and National Security, 23 March 1983, in *Public Papers of the Presidents of the United States: Ronald Reagan, 1983* (in two books), Book I (1 January to 1 July 1983), pp. 437-443, Government Printing Office, Washington, DC, 1984.

Red visitors cause rumpus, *Life*, v. 26, No. 14, pp. 39-43, April 4, 1949.

E. C. Roelof, Energetic neutral atom image of a storm-time ring current, *Geophysical Review Letters*, v. 14, n. 6, pp. 652-655, 1987; doi: 10.1029/GL014i006p00652.

H. Rosenbauer, *Direct measurement of the parameters of the interstellar gas*, Proposal, Max-Panck-Institut fuer Physik und Astrophysik, Muenchen, September 10, 1972.

K. C. Ruffner (editor), *CORONA: America's First Satellite Program*, Center for the Study of Intelligence, Central Intelligence Agency, Washington, DC, 1995.

E. Rutherford, J. Chadwick, and C. D. Ellis, *Radiation from Radioactive Substances*, MacMillan Company, New York, 1930.

A. Schnitzer, Early image-intensifier tube structures, in L. M. Biberman (editor), *Electro-Optical Imaging: System Performance and Modeling*, pp. (4-1)-(4-15), SPIE Press, Bellingham, Wash., 2000.

B. A. Schriever, Our five arsenals of peace, *Space Age*, v. 2, n. 1, Nov. 1959, pp. 36–38.

Scientists protesting (Protestuyut uchenye), Sotsialisticheskaya Industriya, No. 28 (5919), p. 2, February 3, 1989 (in Russian).

Senate Resolution 150, 109th Congress, 1st Session, The Senate of The United States,

May 19, 2005.

V. P. Shalimov, Nachalo nachal (The beginning of the beginning), pp. 24-36, in *Obratnyi Otschet Vremeni. 40 Let Institutu Kosmicheskikh Isslediovanii Rossiiskoi Akademii Nauk (Reverse Countdown of Time. 40 Years of the Space Research Institute of the Russian Academy of Sciences)*, Space Research Institute IKI, Moscow, 2006 (in Russian).

A. A. Shchuka, *Fiztekh i Fiztekhi (Fiztekh and Fiztekh People)*, 5th edition, Fizmatkniga, Moscow, 2012 (in Russian).

L. Silberman, *Tah v. Global Witness*, U.S. Court of Appeals, District of Columbia Circuit, March 19, 2021.

W. B. Smith, *My Three Years in Moscow*, J.B. Lippincott Co., New York, 1950.

M. F. Smith, F. A. Herrero, M. Hesse, D. N. Baker, P. A. Bochsler, P. Wurz, H. Balsiger, S. Chakrabarti, G. Erickson, D. M. Cotton, T. S. Stephen, C. A. J. Jamar, J.-C. Gerard, S. A. Fuselier, A. G. Ghielmetti, S. B. Mende, W. K. Peterson, E. G. Shelley, R. R. Vondrak, D. L. Gallagher, T. E. Moore, C. Pollock, R. Arnoldy, M. Lockwood, R. Gladstone, High-latitude ion transport and energetic explorer (HI-LITE): a mission to investigate ion outflow from the high-latitude ionosphere, in *Instrumentation for Magnetospheric Imagery II*, Proceedings SPIE 2008, pp. 40-56, 1993; doi: 10.1117/12.147643.

A. Smith, Six alternate scenarios that might have reshaped the Trojan engineer-iverse, *USC Viterbi Magazine*, pp. 46-51, Fall 2021; electronic version at https://magazine.viterbi.usc.edu/fall-2021/features/the-great-what-ifs-of-usc-engineering/, accessed on October 16, 2021.

C. P. Snow, *The Two Cultures and the Scientific Revolution*, Martino Publishing, Mansfield Center, CT, 2013.

A. Solzhenitsyn, *The Gulag Archipelago 1918-1956*, Harper & Row, New York, 1973.

G. Sosin, *Sparks of Liberty. An Insider's Memoir of Radio Liberty*, The Pennsylvania State University Press, University Park, Pa., 1999.

M. W. Sowers, G. Hand, and C. M. Rush, Jamming to the HF broadcasting service, *IEEE Transactions on Broadcasting*, v. 34, n. 2, pp. 109-114, 1988.

Srednie zarplaty v Rossii i SSSR s 1897 po 2010 gody (Average salaries in Russia and USSR from 1897 to 2010), Analytical Club, School of Analytical Analysis and Management, http://analysisclub.ru/index.php?page=schiller&art=2757 (in Russian); accessed May 24, 2021.

I. E. Tamm, Teoriya magnitnogo termoyadernogo reaktora, Chast' I (Theory of magnetic thermonuclear reactor, Part I), pp. 3-19, in *Fizika Plazmy i Problema Upravlyaemykh Termoyadernykh Reaktsii (Plasma Physics and the Problem of Controlled Thermonuclear Reactions)*, Publishing House of the USSR Academy of Sciences, Moscow, 1958 (in Russian). English translation: I. E. Tamm, Theory of magnetic thermonuclear reactor (Part I), pp. 1-20, in *Plasma Physics and the Problem of Controlled Thermonuclear Reactions*, Pergamon Press, New York, 1961.

A. Timberlake, *A Reference Grammar of Russian*, Cambridge University Press, New York, 2004.

J. G. Timothy (Principal Investigator), The development and test of ultra-large-format multi-anode microchannel array detector systems, *Progress Report for NASA Grant NAGW-551 for the period 1 June through 30 November 1984*, Center for Space Science and Astrophysics, Stanford University, California.

J. G. Timothy, Electronic readout systems for microchannel plates, *IEEE Transactions on Nuclear Science*, v. 32, is. 1, pp. 427-432, 1985; doi: 10.1109/TNS.1985.4336868.

G. M. Tovmasyan, Yu. M. Khodzhayants, M. N. Krmoyan, A. L. Kashin, A. Z. Zakharyan, R. Kh. Oganesyan, M. A. Mkrtchyan, G. G. Tovmasyan, V. V. Butov, Yu. V. Romanenko, A. I. Laveikin, A. P. Aleksandrov, and D. Huguenin, The Glazar orbiting ultraviolet telescope, *Soviet Astronomy Letters*, v. 14, n. 2, pp. 123-124, 1988.

D. E. Voss and S. A. Cohen, Low-energy neutral atom spectrometer, *Review of Scientific Instruments*, v. 53, pp. 1696-1708, 1982; doi: 10.1063/1.1136873.

J. L. Wiza, Microchannel plate detectors, *Nuclear Instruments and Methods*, v. 162, pp. 587-601, 1979.

P. Wolfowitz, Statesman of the century, *Wall Street Journal*, v. CCLXXVII, No. 32, p. A19, February 9, 2021.

P. Wurz, letter to the author, dated January 15, 1993.

Xinhua News Agency, Space will see Communist loyalty: Chinese astronaut, October 21, 2017; http://www.xinhuanet.com/english/2017-10/21/c_136695873.htm; accessed on November 6, 2020.

V. Yasmann, n.d.; Victor J. Yasmann papers, Collection Number: 2014C8, Hoover Institution Library and Archives, https://oac.cdlib.org/findaid/ark:/13030/c8pc3426/; accessed January 15, 2021.

L. M. Zelenyi (general editor), *Institut Kosmicheskikh Issledovanii Rossiiskoi Akademii Nauk. 50 let* (*Space Research Institute of the Russian Academy of Sciences. 50 Years*), Space Research Institute IKI, Moscow, 2015 (in Russian).

V. K. Zworykin, G. A. Morton, and L. Malter, The secondary emission multiplier—a new electronic device, *Proceedings of the Institute of Radio Engineers*, v. 24, n. 3, pp. 351-375, 1936.

Index

Numbers

8K63. See SS-4
8K64. See SS-7
8Zh38 12. See R-2
1979 List 87, 164, 165, 167, 171, 174, 193, 195, 203

A

A-4, ballistic missile 68
Abkhazia 147, 148
academic freedom 113
Advanced X-ray Astrophysics Facility, AXAF 169
Aeroflot ix
Aeronomical Service (Service d'Aéronomie) 224
Afanasiev, Victor 101, 103, 107, 117
AFL-CIO 137
Africa 81
Agency for the Global Media, U.S. 134, 136, 143, 144, 146
AIAA 250
Ainbund, Mikhail R. 60, 87-89, 92, 95
Akademgorodok 2, 3
Akademik Sergei Korolev, ship 124, 125
Akmolinsk 140
Albania 141
ALZhIR prison camp 140

American Geophysical Union 166, 182, 212
American Military Government 144
American Physical Society 31
AMOLF. See FOM Institute of Atomic and Molecular Physics
Andropov, Yuri V. 40, 115, 137, 190
Angarov, Vadim 39
Anger, Hal 92, 96
antisemitism 4, 89, 112
APL. See Applied Physics Laboratory
Applied Physics Laboratory 99, 157, 211-213, 220, 224, 225
Arctic Circle 120, 121
Arctic Ocean 122
Armenia 108, 116, 122, 123, 227, 229, 230, 238
Armenian Academy of Sciences 108
Artsakh 227-230
aspirantura 3, 4
Astana 140
ASTRID, space mission 233
avtoreferat 67, 68
Axford, Ian 180, 200, 201, 213
Azerbaijan 227, 228, 230, 238

B

Bad-Honnef 178
Baikonur Cosmodrome ix, 237

301

index

Baku 227, 228
Balebanov, Vyacheslav M. 95
Balega, Yuri 117
Baltic Way 242, 243
Banaszkiewicz, Marek 214
Barabash, Stas 231
Baranov, Vladimir B. 33-35, 40, 42, 44, 53, 123, 147, 157, 158, 172, 173, 175, 179, 197-199, 214, 215, 219, 220, 223-225, 227, 238, 239
Barcelona 133, 134
Baspik. See Gran, plant
BBC. See British Broadcasting Corporation
Bell Telephone Laboratories 54
Belotserkovsky, Oleg M. 30
Belyaevo, metro station 20, 21
Bendix Corporation 55, 63
Berlin 177
Berlin Wall 177
Bernstein, Bill 159
Bertaux, Jean-Loup 224
Beryoza, field position 14, 15
BESM-6 45
Bishkek 39
Bleszynski, Stanislaw 214
Blum, Peter 117, 176, 178, 221, 226
Bolshoi Zelenchuk 105
Bowyer, Stuart 96, 123
Brezhnev, Leonid I. 175, 176, 190
British Broadcasting Corporation 130, 132, 144, 145
British Communist Party 148
Broadcasting Corporation of China 136
Brookhaven National Laboratory 194, 195
BTA-6 101-105
Bujak, Zbigniew 206
Bulgaria 138, 223
Bush Derangement Syndrome 111
Bush, George W. 111
Byurakan Astronomical Observatory 108, 116, 226, 227
Bzowski, Maciej (Maciek) 214, 218, 220

C

CAMAC 71, 92, 93, 98
Canada 122
Cassini, space mission 233
Catalonia 133, 134
Cavendish Laboratory 8
CBK. See Space Research Center, CBK
CEM. See channel electron multiplier
Central Committee, CPSU 6, 7, 18, 132
Chalov, Sergei V. 157, 158
Chandra X-ray Observatory 169
channel electron multiplier 55, 57, 60, 62, 63, 186
charge-coupled device, CCD 62
Chebrikov, Victor M. 132, 144, 145
Checkpoint Charlie 177
Cheng, Andy 213
Cherenkov, Pavel 155
Chernyaev, Anatoly 41, 137, 138
Chernyshev, Georgii P. 127
Chicago xii
Chief Directorate of Space Assets 23, 24
China 16, 40, 41, 48, 112, 113, 130, 136, 139, 145
Chinese Communist Party 39, 41
Churchill, Winston 149-151
CIA 15, 26, 61, 130, 133
COCOM 128
Cohen, Sam 63, 64
Cold War xiii, 112, 122, 128, 129, 133, 175, 177, 219, 251
Committee for State Security, KGB 40, 41, 80, 81, 109, 113, 114, 119, 127, 128, 132, 137, 147, 148, 229
Committee on Space Research. See COSPAR
Communist Party of China 48
Communist Party of the Soviet Union vii, 2, 6, 7, 15, 18, 19, 34, 38, 39, 40-43, 47, 49, 78, 82, 89, 100, 129, 132, 133, 137-139, 145, 148, 149, 175, 198, 235, 239, 242
COSPAR 201, 213, 231, 232
Costa Brava 134

index

Council of Ministers, USSR 8, 18, 19, 148
CPSU. See Communist Party of the Soviet Union
CRRES, space experiment 233
Cuba 112, 139
Culham, England 174
Curtis, Charles 182, 183
Czechoslovakia 42, 81, 138, 144
Czech Republic 145, 147

D

Darmstadt 168
Day of Cosmonautics 127
Demchenkova. See Kozochkina (Demchenkova), Alla
Democratic Party 129, 136
Department No. 4 43
Department No. 18. See Department of Space Gas Dynamics
Department of Space Gas Dynamics xi, xii, 4, 33-35, 38-40, 42, 43, 101, 108, 124, 157, 158, 175, 214, 223, 235, 236, 238
Deutsche Welle. See German Wave
diffraction filter 167
Discovery, Space Shuttle 188
Dnepropetrovsk 12, 13
Dolgopa. See Dolgoprudny
Dolgoprudny 8, 9, 10, 11, 31
D.Sc. degree 67, 118
Dubna 168
Duga, radar 132
Duke of Wellington 253
Dulles, Allen 133, 138
Dzerzhinsky, Felix 114

E

Eastern Europe 131, 216, 222
École Polytechnique 6
Electron, space launcher 244
ENA imaging 53, 66, 87, 154-156, 159, 160, 162, 169, 202, 205, 209, 210, 213, 224
Energia, Rocket and Space Corporation 74
Engels, Friedrich 41

Englischer Garten 142, 143, 145
ES-1040 45
Estonia 242, 243
ethanol 72
ethyl alcohol 72
Etkin, Valentin S. 26
European Space Agency 189
Evlanov, Evgenii N. 36-39, 43, 73, 74, 81

F

Faculty of Aerophysics and Space Research 1, 16, 17, 29, 30
Faculty of Problems of Physics and Energetics 29, 30
Fahr, Hans 35, 49, 117, 175, 176, 178, 186, 198, 214-217, 219, 223, 226, 227
FAKI. See Faculty of Aerophysics and Space Research
Farragut, David 60
Federal Republic of Germany 136, 176, 177, 179, 180, 184, 219
Feldman, William (Bill) 186
fellow traveler 41, 109, 137, 139, 243
Finland 119, 122, 223
Firmani, Claudio 103
Fiztekh. See Moscow Physical-Technical Institute
Flerov, Georgii N. 168
foil, thin 64, 87, 161-164, 174, 190, 193-195, 199, 200
FOM Institute of Atomic and Molecular Physics 45, 47-49, 87
France 67, 139, 142, 148, 223, 224
Frank, Ilya 155
Freedom of Information Act, FOIA 61
Frunze 39. See Bishkek

G

Gagarin, Yuri 127
Galeev, Al'bert A, xii, 18, 26, 27, 36
Galperin, Yuri I. 36
Gamow, George ix, 48, 49, 122
Garching 186, 223

index

GAS-1 199, 201
GAS-2 199, 201
GAS-3 199, 201
GAS-E 199, 201
GAS experiment 65, 189, 196-203, 209, 214, 235
GAS, Ulysses 65
Gdansk 205-207, 238, 242
GDR. See German Democratic Republic
Geiger, Johannes 165
Georgia 114, 141
Georgian Academy of Sciences 98, 115
GEOTAIL, space mission 233
German Democratic Republic 176, 177
German Wave 130, 132, 144, 145
Germany xiv, 16, 38, 40, 49, 64, 67, 81, 110, 113, 118, 130, 133, 135, 136, 138, 142, 143, 146, 147, 168, 175-180, 184, 200, 201, 210, 213, 217, 219, 223
Giotto, space mission 38
Gissar Range 120
Glazar, space telescope 108
Gloeckler, George 202, 203, 223
Glukhovskoi, Boris M. 101
Goebbels, Joseph 41
Goertz, Christoph 166
Goettingen 180
Goodrich, Goerge W. 54, 55
Gorbachev, Mikhail S. vii, 5, 40, 44, 49, 137, 229, 238, 243
Gorky, Maxim 42
Gorn, Lev S. 87, 89, 90, 92, 98, 199
Gott, Yuri V. 54, 57, 67, 88, 193
Gran, plant 85, 88
Great Britain 50, 139, 142, 223, 253
Gringauz, Konstantin I. 36, 89, 90, 123, 126, 184, 238
Grizzard, Lewis vii, 68
Grzedzielski, Stanislaw (Stan) 117, 175, 176, 179, 182, 196, 198-200, 203, 204, 207, 210, 213-216, 218, 221, 226, 227, 231
Guenzburg 219

GUKOS. See Chief Directorate of Space Assets
Gulag 44, 121, 140
Gvishiani, Dzhermen 148
Gvishiani-Kosygina, Lyudmila 148
Gwiazda, Andrzej 206

H

Halley, comet 38, 39, 182, 197, 204
Heidelberg 38
heliopause 160, 227
heliosphere xi, 53, 156, 157, 159, 160, 172, 173, 176, 180, 182, 203, 209, 211, 215, 216, 218, 219, 223, 224, 247
Hlond, Marek 199
Holodomor 139
Hoover Institution 207
Hsieh, K.C. (Johnny) 181-183, 213, 220
Hungary 16

I

IBEX. See Interstellar Boundary Explorer
IBM 45, 46
IBM System/360 45
identity politics 3, 48
IMAGE, space mission 165, 201, 209, 210, 233
IMP 7, space mission 211
IMP 8, space mission 211
Institute for Problems in Mechanics 35, 38, 158, 223, 236, 238, 239, 243
Institute of Atomic Energy 30, 67, 193, 199
Institute of Physics, U.K. xvi, 31, 49, 97, 186, 192
intercontinental ballistic missile 12, 132
intermediate-range ballistic missile 12, 13, 15
Interstellar Boundary Explorer xii, 59, 64, 65, 160, 162, 165, 167, 173, 202, 203, 209, 210, 218, 224, 225
interstellar boundary, solar system xii, 64, 65, 151, 154-157, 159-162, 165, 167, 171, 173, 196, 199-203, 205, 209, 210, 218, 224, 225

index

interstellar helium 52, 53, 57-59, 62-65, 157, 161, 176, 180, 181, 186, 188, 189, 198, 199, 212, 214
Interstellar Pathfinder, space mission 202, 203
Interstellar Probe, space mission 65
Ioffe, Abram F. 45
Ioffe Physical-Technical Institute. See Leningrad Physical-Technical Institute
IOP. See Institute of Physics, U.K.
IPM. See Institute for Problems in Mechanics
Iran 112
ISEE 1, space mission 211, 212
Ishlinsky, Alexander Yu. 236
Israel 112, 130, 132, 144, 145
item number five, the 3, 4
Izmodenov, Vlad (Vladislav) xi, xii, 157, 158

J

jamming 130-134, 142, 145
Jankowski, Henryk 206
Japan 128, 223
Jaruzelski, Wojciech 204
Jet Propulsion Laboratory 166, 188, 220
Joint European Torus, JET 174
Joint Institute for Nuclear Research 168
Judge, Darrell 35, 220-222, 245, 247, 249
Jupiter 196, 199, 211, 214

K

kafedra of space physics 26, 29, 30
Kalinin, Aleksandr (Sasha) P. xv, 36, 37, 47, 59-62, 65, 72, 74, 86, 108, 190
Kaliningrad. See Podlipki
Kaluzhskaya, metro station 20-22
Kapitsa, Petr L. 6, 8, 29, 30
Karabakh. See Artsakh
Karabakh Committee 229
Karelia 119
Katlenburg-Lindau 180, 182, 201, 213, 214
Kazakhstan 139, 140, 237, 238
Kennan, George 139
KGB. See Committee for State Security, KGB
KH-4, Corona xiv, 20-22

KH-9, Hexagon xiv, 10, 11, 14, 25, 104-106
Khazanov, Boris I. 87, 89, 90, 92, 98, 199
Khodarev, Yulii K. 26
Khristianovich, Sergei A. 6, 8
Khromov, Vladimir (Volodya) N. 36, 38, 42, 47, 65, 66
Khrushchev, Nikita S. 12, 40
Kiev xvi, 101
Kiruna 231
Kistemaker, Jaap (Jacob) 45, 48
Kol Israel. See Voice of Israel
Korolev, Sergei P. 74
Kosygin, Aleksei N. 18, 148
Kozochkina (Demchenkova), Alla 37, 86, 101
Kraft durch Freude 118
Krivitsky, Walter 138
Kuban River 105
Kubetsky, Leonid 54
Kurt, Vladimir G. 36, 51-53, 67
Kvant, astrophysics module, Mir 108
Kyrgyzstan 39

L

Lallement, Rosine 224
Lampertheim 133, 135, 136
Lampton, Mike 55, 96, 98, 99, 103, 114, 115, 123, 196
Landsat 5
Las Cruces 248
Latin America 81
Latvia 15, 242, 243
LAX, airport x, 245, 247
Lay, Guenter 219
Lazarus, Alan 224
Lebedev, Mikhail 220
Leipzig 81
LENA, ENA instrument 167
Leningrad 45, 67, 87, 98, 107, 114, 116, 118, 126, 174
Leningrad Physical-Technical Institute 45, 67, 98, 174
Lenin Shipyard 206
Lenin, Vladimir I. viii, 41, 42, 110, 141, 206

305

index

Leonas, Vladas B.   xi, 1-4, 31, 33-38, 40, 42, 45, 47, 51, 52, 57, 63, 64, 69, 76, 86, 87, 95, 97, 98, 116, 118, 123, 127, 155-157, 159, 164, 166, 168, 173-176, 179, 190, 195, 197-199, 213, 214, 219, 220, 223-225, 236, 238
L'Humanité   148
Library of Foreign Languages   148, 149
Liechtenstein, Vitaly Kh.   193-195, 199
Ligachev, Yegor K.   132, 144, 145
light table   250
Limbaugh, Rush   243
liquid nitrogen   78, 81
LISM.   See local interstellar medium (LISM)
Lithuania   107, 242, 243
Local Interstellar Medium Explorer (LIME)   224
local interstellar medium (LISM)   52, 65, 157, 158, 160, 161, 165, 171, 172, 203, 219, 223, 224
Lockheed-Martin   157
Los, Joop   45, 47, 48
Lundin, Rickard   231
Luzhniki, sports arena   20

M

Macek, Wieslaw   214, 215
Madrid   130
magnetosphere   xi, 87, 99, 153-157, 159, 161, 196, 199, 210-213, 224
Malama, Yuri G.   157, 158, 173, 220
Managadze, Georgii G.   36
Mao Zedong (Tse-tung)   41
Mars-3, space mission   159
Massachusetts Institute of Technology   170, 224
MASTIF   173, 189-191
Max Planck Institute   38, 180, 181, 186, 199, 201, 223
Max Planck Institute for Aeronomy   180, 184, 185, 199-201, 213, 214
Max Planck Institute for Extraterrestrial Physics   186
Max Planck Institute for Nuclear Physics   38

Max Planck Institute of Physics and Astrophysics   186
Max Planck Society   xiv
Mazovia Province   206, 207
Mazovia Weekly (Tygodnik Mazowsze)   207
McComas, Dave   165, 170, 202, 203, 218
McMullin, Don   247, 248
MCP.   See microchannel plate
MENA, ENA instrument   165, 170
methanol   72
Mexico city   103
MFTI.   See Moscow Physical-Technical Institute
MGU.   See Moscow State University
microchannel plate   45, 47, 55-57, 62, 63, 66, 85-95, 98-101, 103, 108, 123, 162, 164, 165, 168, 190, 193, 200
Middle East   81
Military-Industrial Commission, USSR   18
Mineralnye Vody   226
Ministry of Defense, USSR   18, 23, 24
Ministry of General Machine Building   18, 19, 69
Ministry of Interior   76
Ministry of Middle Machine Building   89
Mir, space station   108
missile defense   115, 251
Mitchell, Don   220
Mitrofanov, Alexander   168, 169
Moebius, Eberhard   202, 223, 224
Molotov-Ribbentrop pact, 1939   242
MOM.   See Ministry of General Machine Building
Monte Carlo technique   45, 173
Morning Star   148
Morozov, Victor (Vitya)   37, 61, 86, 164, 165, 189, 195
Morse code   131
Moscow Electronic Valve Plant   101
Moscow Helsinki Group   113
Moscow Institute of Physics and Technology
   1.   See also Moscow Physical-Technical Institute

index

Moscow Physical-Technical Institute 1-3, 5-12, 14-17, 26, 29, 30, 31, 37, 58, 66, 101, 149
Moscow River 10, 20
Moscow State University 8, 20, 158, 172, 220
MPAe. See Max Planck Institute for Aeronomy
Mukhin, Lev M. 36
Munich 136, 139, 142-147, 186, 219, 223
Munich agreement 139, 144
Murmansk 119
Murphy, Robert 139, 142
Mussolini, Benito 41

N

Nagorno-Karabakh 227, 228, 230. See also Artsakh
Nalchik 101
National Institute of Standards and Technology, NIST 170
National Socialist German Workers' Party, NSDAP 40
NATO 219
negative ion 162-167
Netherlands vii, 45, 48
Newton, Isaac 8
night-vision device 55, 56, 62
nitric acid 13
nitrogen tetroxide 13
Nizhnii Arkhyz 103, 105
Nobel Peace Prize 207
Nobel Prize in literature 138, 149
Nobel Prize in physics 29, 30, 113, 155
nomenklatura 2, 116, 124
North Korea 145
Novodachnaya, platform (station) 11
Novosibirsk 2, 3
Novye Cheremushki, metro station 20, 21
NTF. See nuclear track filter
nuclear track filter 168, 169, 200
Nur-Sultan 140

O

Oberwiesenfeld, airport 142, 144
Obraztsov, Ivan F. 30
Obrucheva, street 20, 21, 22
Odessa 124, 125, 126
Odessa Polytechnic Institute 124
Ogawa, Howard 220, 222, 245, 247
Ordzhonikidze 85, 88
Orlov, Yuri F. 40, 49, 113, 243
Orwell, George 44, 127, 129, 139, 204, 253
Ostrov 14, 15

P

Pamir-Alay 120
Pan American x, 245
passive corpuscular diagnostics 155, 156, 174
passport viii, 3, 148
Patrice Lumumba University 81, 82
People's Republic of China 16, 40, 48, 130, 139, 145
Pervomaisk, Ukraine 13
Petrov, Georgii I. 18, 23, 33, 34, 35, 40, 67, 155, 174, 196, 213, 236
Petrukovich, Anatoly 18
Ph.D. degree xiii, xiv, 2, 15, 27, 29, 52, 66-68, 118, 193, 239
Phobos 123, 195, 196, 197, 199
Phobos 1, space mission 196
Phobos 2, space mission 196
Physical Institute, FIAN 168, 169
Physics Institute, Vilnius 107
Pierce, John R. 54
Pioneer 10, space mission 246, 247
Plant No. 632 101
Playa de Pals, Spain 133, 135, 136
Podgorny, Igor M. 37
Podlipki ix, x, 4, 5, 74
Poland viii, 44, 48, 71, 132, 138, 200, 203, 204, 206-208, 210, 213, 214, 218-220, 223, 231
POLAR, space mission 233
Polish Academy of Sciences 175, 179, 213, 220, 232

index

Politburo 132, 137, 144, 145, 175
Portugal 133
position-sensitive detector xii, 35, 45, 47, 49, 55, 56, 62, 63, 66, 74, 83, 87-89, 91, 93-101, 103, 108, 123, 128, 151, 162, 168, 236
Prague 81, 145, 147
Pravda 6, 7, 115, 129
Pribory i Tekhnika Eksperimenta 59, 60, 86, 89, 95, 97, 128
Princeton Plasma Physics Laboratory 63
Princeton University 63, 218
Profsoyuznaya, street 20-24
Prokhorov, Alexander 29, 30
Proton, space launcher 196
Proudhon, Pierre-Joseph 41
PTE. See Pribory i Tekhnika Eksperimenta
Pulkovo, airport 114
PUMA, space instrument 38, 39, 197

R

R-2 12
R-12. See SS-4
R-16. See SS-7
Radio Corporation of America 54
Radio Free Europe 130, 133, 134, 136, 142-147
Radio Korea 145
Radio Liberty 130, 132-136, 142-147
Radio Peking [Beijing] 145
Radziejowice 220
RATAN-600 103, 106, 107, 117, 226
Ratkiewicz, Romana (Roma) 218, 219
Reagan, Ronald 115, 141, 210, 243, 253
Regime Department, IKI 49
Reichsarbeitsdienst, Reich Labor Service 16
Relikt-2, space mission 199, 201, 202
residence permit 3, 4
RFE. See Radio Free Europe
Richter, Arne 213
ring road, Moscow 5, 10, 20, 21
RL. See Radio Liberty

Roelof, Edmond (Ed) 157, 211, 212, 220, 224, 225
Rolland, Romain 138
Rosenbauer, Helmut 64, 180-182, 184-189, 196, 197, 199-201, 213, 214, 226, 227, 238, 241
Rubel, Marek 218
Rucinski, Daniel 214, 215
Ruderman, Mikhail (Misha) S. 157, 158, 172
Rutherford, Ernest 6, 8, 30, 165

S

Saddler. See SS-7
Sagdeev, Roald Z. 3, 4, 17, 18, 27-31, 33, 35, 38, 67, 82, 127, 175, 179, 181, 196-198, 200, 201, 210, 235-237, 239, 242
Saint Petersburg. See Leningrad
Sakharov, Andrei 239, 240, 242
Sandal. See SS-4
SAO. See Special Astrophysical Observatory
Saturn 196, 199, 211
Schattenburg, Mark 170
Schiphol, airport vii, ix, x, 245
Schriever, Bernard 111, 112
Science magazine xii, 65, 173
secondary electron multiplier 53, 54, 57, 58, 62, 88, 161
secondary ion mass spectrometry 173, 189, 190
Secret Department, IKI 49
SEM. See secondary electron multiplier
Serock 231, 232
Shapiro, Vitalii D. 26
Sheremetyevo, airport vii, 176, 195
Shusha 227, 229
Siberia 121, 205
Sibling. See SS-2
SNIIP 87-90, 95, 97, 98, 199
Snow, Charles P. 111
Solidarity, trade union 48, 132, 204-207, 208
Solzhenitsyn, Alexander 40, 44, 204
South America 251

index

Space Astrophysics Group 96, 98, 123
Space Gas Dynamics Conference 178, 212, 221, 223, 224, 230
Spacelab, Space Shuttle 98
Space Research Center, CBK 44, 117, 175, 176, 179, 196, 198-200, 203, 204, 207, 213-215, 218, 219, 232
Space Sciences Center, USC 245-248
Space Sciences Laboratory, Berkeley 95, 98, 99
Space Shuttle 98, 115, 188
Spain 130, 133-136, 148
Special Astrophysical Observatory 101-103, 107, 108, 116, 117, 226
SS-2 12
SS-4 12-16
SS-7 12
Stalin, Joseph (iosif) V. 8, 41, 110, 138, 141, 168
Stasi, Ministry for State Security, GDR 49
St. Bridget's Church 206
steklyashka (glassy house) 21, 22, 23
Stepanakert 227, 228
Stevens Institute of Technology 166
Stockholm 16
Strategic Rocket Forces, USSR 12-15
Sumgait 227-230
Sunyaev, Rashid A. 36, 123
surface conversion 165-167
Sweden 128
Swedish Institute of Space Physics 231
Swedish Military Museum 16
Synchrotron Ultraviolet Radiation Facility 170

T

Taiwan 135, 136
Tajikistan 120
Tamm, Igor E. 155
Tbilisi 98, 114, 115, 141
tenure 26, 113
termination shock 160
Three Crosses (Trzy Krzyże), monument 206
Tirana 141

Tolstoi, Ivan 145, 147
Troitsk 30
Trump, Donald J. 111
TRW Systems 159
Tucson 181-183
tunneling transition viii, ix, x, 221, 245
Turkey 122, 230
TWINS, space mission 165, 170, 209, 210, 233
Tygodnik Mazowsze. See Mazowia Weekly
Tyuratam Missile Test Range ix, 237

U

Ukraine 12, 13, 124, 139
Ulysses, space mission 59, 65, 188, 189, 196, 199, 214, 233
Umansky, Sergei V. 37, 38, 74, 79
Universita 17. listopadu 81
University of Arizona 166, 181, 183, 213, 220
University of Bern 167
University of Bonn 35, 117, 175, 176, 178, 186, 213-215, 219, 221
University of California, Berkeley 95, 99
University of Cambridge 8, 111
University of Southern California xi, 35, 157, 159, 166, 170, 171, 182, 189, 220, 222, 245, 246, 248, 249, 251, 252
USAGM. See Agency for the Global Media, U.S.
U.S. Congress 138
U.S. Senate 133, 139

V

V-2, ballistic missile 68
Vaisberg, Oleg L. 36, 89, 90, 124, 126, 159
Van Allen, Jim 166
Vandenberg Air Force Base 209
Van'yan, Leonid L. 29, 31
Vasyukov, Sergei 43
Vega 1, space mission 204
Vega, space mission 38, 39, 182
Velikhov, Evgeny M. 30
VEU-1 54, 88

309

index

VEU-6 57, 59, 60, 86, 88
VEU-7 88, 100
Vilnius 107
Vladikavkaz 85
VNII Elektron 87-89, 95, 97
VNIIT 107, 108
VOA. See Voice of America
Voice of America 130, 132, 136, 144, 145
Voice of Israel 130, 132, 144, 145
Volga River 10, 119
Voss, Hank 64, 157
Voyager (1 and 2), space missions 160, 211, 224
Vremya, TV newscast program 137, 142

W

Walesa, Lech 205, 206, 207
Warsaw viii, 44, 48, 117, 175, 176, 179, 182, 196, 198-200, 203-208, 213-215, 218, 220, 231, 232
wedge-and-strip collector 95, 96, 97, 98
Werner, Otto 165
White Sands Missile Range 247, 248
Wiley, William C. 54, 55
Williams, Don 213
World Federation of Trade Unions 138
World Peace Council 138
World War II 6, 68, 139, 142, 144, 168, 205
Wurz, Peter 167

Y

Yang Liwei, taikonaut 43
Yasmann, Victor 145, 147
Yeltsin, Boris N. 39
Yerevan 227, 229, 230
Yuzhnoe Design Bureau 12, 13

Z

Zakopane 218, 219
Zelenchukskaya 103, 105, 107
Zelenyi, Lev M. xiii, 18, 27, 119
Ziemkiewicz, Joanna 219
Zodiak, space experiment 219

Zolotukhin, Valerii G. 26
ZOMO 206, 207
Zubkov, Boris (Borya) 37-39, 74, 81, 82, 236
Zworykin, Zworykin 54

Other books by Mike Gruntman

Mike Gruntman

Blazing the Trail: The Early History of Spacecraft and Rocketry

AIAA, Reston, Va., 2004

Winner of the Luigi Napolitano Award (2006) from the International Academy of Astronautics

ISBN 1-56347-705-X
475 pages with 340 figures

ISBN 978-1-56347-705-8
Bibliography: 250+ references
Index: 2750+ entries, including 650 individuals

Mike Gruntman

From Astronautics to Cosmonautics.
Space Pioneers Robert Esnault-Pelterie and Ary Sternfeld.

Booksurge, North Charleston, S.C., 2007

Nominated for the Eugene M. Emme Astronautical Literature Award (2007) of the American Astronautical Society

ISBN 978-1-4196-7085-5
ASIN: B002E19WDO (Kindle edition)
84 pages with 24 photographs

ISBN 1-4196-7085-5 (paperback)

Bibliography: 75 references
Index: 230+ entries

Mike Gruntman

Enemy Amongst Trojans. A Soviet Spy at USC.

Figueroa Press, Los Angeles, Calif., 2010

ISBN-10: 1-932800-74-3
88 pages with 12 figures

ISBN-13: 978-1-932800-74-6
Bibliography: 94 references
Index: 90+ entries

Mike Gruntman

Intercept 1961. The Birth of Soviet Missile Defense.

AIAA, Reston, Va., 2015

ISBN 978-1-62410-349-0 (print)
eISBN: 978-1-62410-350-6 (pdf at http://arc.aiaa.org)
330 pages with 120+ figures

Bibliography: 200+ references
Index: 800+ entries